The Principles of Modern Biology

One Approach
to Ecology

Arnold Darlington, M.A., M.B.O.U., F.I.Biol.
Alison Leadley Brown, M.A., F.I.Biol.

LONGMAN

LONGMAN GROUP LIMITED
London
*Associated companies, branches and representatives
throughout the world*

First published 1975
ISBN 0 582 32321 5

Printed in Great Britain by
Whitstable Litho, Whitstable, Kent

Titles in the series

Molecules and Cells
Genes, Chromosomes and Evolution
Statistics for Biology
Biology of the Individual
The Working Plant
An Introduction to Functional Systems in Animals

Acknowledgements

The authors and publishers are grateful to the
following for permission to reproduce photographs:
Tony Haynes, pages 7, 8, 9 and 10 and Keystone Press
Agency, page 157.

Cover photograph: Imitor.

The authors are grateful to Peter Stebbing for the care
and research he has devoted to the drawing of
organisms for this book.

Contents

	Page
FOREWORD	vi
INTRODUCTION	viii

1 THE ORGANISM AND ITS ENVIRONMENT — 1

A woodlouse, *Oniscus asellus* — 1
Problems imposed by evolution — 1
Identification — 1
Oniscus as a living organism — 2
Behaviour under natural conditions — 10
Water loss — 11
Water gain — 13
Temperature — 13
Distribution — 14
Experimental work — 15
Key to the common genera of British woodlice — 21

2 THE ORGANISM AND ITS ENVIRONMENT — 25

Creeping buttercup, *Ranunculus repens* — 25
Distribution and morphology — 25
Reproduction by flavering — 28
Dispersal of sexual propagules — 30
Dispersal of asexual propagules — 35
Effects of light — 43
Relationship between light and growth — 44
Effects of water — 44
Relationship between substrate water and buttercups — 48
Some conclusions — 53
Key to British plants with buttercup-like flowers — 54

3 THE HABITAT — 57

What is a habitat? — 57
More restricted habitats — 57
The less restricted habitats — 60

Contents

		Page
4	RELATIONSHIPS AND INTERDEPENDENCE	64
	The concept of community	64
	Close relationships	64
5	ENERGETICS	73
	Origin of the energy flow	73
	Trophic relationships within the ecosystem	73
	Food chains and food webs	74
	Ecological pyramids	78
	Energy flow through an ecosystem	80
	Rate of energy transfer	82
	Ecological efficiency	85
6	STABLE SITUATIONS	87
	What is a stable community?	88
	Palaeo–ecology and past successional changes	89
	Climax and subclimax ecosystems	94
	The grassland subclimax	96
	The ecological niche	111
	Autecology and synecology	123
7	POPULATIONS AND CHANGES IN THEIR SURROUNDINGS	125
	Interactions between colonizers	125
	Interrupted succession	129
	Effects of plant succession on animal communities	134
	Effects of plant succession on a single species	136
8	COLLECTING METHODS AND SAMPLING POPULATIONS	144
	Methods of collecting	144
	Some problems of sampling	148
	Estimating numbers	149
	Estimating a population of common toads, Bufo bufo	152
	Quantitative studies of a blue-tit population over-wintering in a town	156
	Estimating bird populations by the alighting method	163

Contents

		Page
9	MAN'S PLACE IN NATURE	165
	The destruction of climaxes and the effects on soil fertility	165
	Weeds and competition	168
	Relationship of plant cover to erosion	169
	Pollution	171
	Pesticides	172
	Some alternatives to the use of pesticides	174
	Chemical pollution of water	176
	Organisms promoted by man	178
10	CONSERVATION OR PRESERVATION?	180
	BIBLIOGRAPHY	187
	GLOSSARY	191
	INDEX	197

Foreword

In recent years, *ecology* has become an emotive word and certainly a fashionable one, used in such a range of contexts and at such a variety of standards that perhaps it should not be included in the title of yet another book without qualification.

How often does one meet the student who, willing initially to apply his mind to the problems of 'ecology', has suffered frustration from the complexities covered by the word and the statistical requirements to elucidate them? He may be someone who was attracted to biology in the first place because the subject had an appeal for him which was aesthetic rather than mathematical. His formative years may have been restricted to traditional exercises: if so, he could find himself at a particular disadvantage in having been concerned mainly with the parts of an organism and not with the organism as a whole. Any living material he handled was provided by the laboratory staff, and any functions it was made to perform were considered more or less in isolation from the requisite conditions for environmental success. To him, a dogfish is a dead body without a tail, a crayfish is another dead body with an incredible number and variety of limbs, the chief feature of both being that they smell strongly of preservative. He has acquired little knowledge of the forms of life sharing his own environment, and still less of where to find them. Yet sound natural history remains what it has always been, the basis of sound ecology.

Such a person, frequently imaginative and intelligent and potentially a good student, merits consideration. We believe that a return to relative simplicity in ecological studies, commensurate with the range inherent in the varied aspects of the subject, is overdue. We are concerned with three ways of achieving this: the investigation of organisms and situations in the student's surroundings, an autecological approach as opposed to that of synecology (both require a knowledge of natural history), and a limited use of statistics. Whereas the importance of quantitative work is recognised, the temptation is set aside of assuming that, merely because a number can be attached to something, greater understanding is necessarily attained thereby.

We advocate a shift in emphasis, but we do not accept a decline in standards.

With these aims in mind, we include materials and methods we ourselves have used at advanced levels of teaching. Where information comes from the work of others, this is stated. Practical examples, from whatever source, suggest attitudes applicable to situations in many parts of the world. But if the phenomena with which they are concerned are considered only in theory or the narrowest of laboratory contexts, they have little meaning when field conditions come to be assessed.

Students who require statistical models and more refined techniques should consult the specialist literature in the bibliography.

Arnold Darlington
Alison Leadley Brown

Introduction

Ecologists account for the occurrence of a particular organism in a particular combination of surrounding conditions, and its exclusion from others, by the interaction of various factors. Those associated with the soil (edaphic factors), the climate, and the effects of other living things upon the organism itself and upon each other (biotic factors) are among the influences which together make up the organism's environment. Attempts by students to evaluate the ecology of an entire region, or that of a single species, frequently become confused through the imprecise and indiscriminate use of such terms as 'habitat', 'environment', 'community' and 'ecosystem'. Comprehension of what is implied arises only from the experience borne of practice. For this reason, the first two chapters are concerned with the detailed examination of one animal, the common woodlouse, *Oniscus asellus*, and one plant, the creeping buttercup, *Ranunculus repens*, in respect of the relationships of each with its own surroundings. Such an examination constitutes an autecological investigation of the species.

Both the organisms selected colonize situations which are more or less definable, although both are widely distributed and often abundant where they occur. Other animals and plants come to mind which might be substituted. It will be apparent that an ecological investigation requires a knowledge of the taxonomy, behaviour and physiology of the species concerned. It is equally obvious that to each species must be applied its particular criteria, since the environmental features determining the success or failure in its surroundings of a woodlouse or a grassland buttercup are not necessarily significant in the case, say, of a freshwater invertebrate or a succulent plant exposed to aridity. The methods applied to the two organisms we have chosen serve as a guide to the sorts of tests adaptable to numerous other cases and underline the fact that a study in depth of the ecology of a particular species carries its own problems. It is the elucidation of these special problems which paves the way for the more complex task of assessing the reactions and interactions of organisms living together in a particular region–in a word, their synecology. Topics which we have considered and which touch organisms both at autecological and synecological levels include the following.

The habitat of any living thing is the place where it lives and from which it can obtain three basic needs, food, shelter and somewhere to breed. Chapter 3 considers the concept of 'habitat' and whether such a summary covers all its features. A so-called habitat may be difficult to envisage as an entity when the organism which occupies it has a relatively unrestricted range.

Although a community can be regarded as comprising organisms sharing a common environment, the complexity of many a community is such as to make for difficulty in assessing the correlation between its components on the one hand and their environmental conditions on the other. Relationships examined in Chapter 4, where one organism provides habitats for others, proffers the reservation that such intimate associations are to be considered as those of degree, since the connection between an apparent parasite and its host may produce results beneficial to both. Energy flows through all ecosystems. Origins of the energy and its passage through the different trophic levels are outlined in Chapter 5, and examples are given of food webs to illustrate the complexity of energy relationships.

Environmental conditions are seldom those of stability. The changes in stability may lead eventually to climax or subclimax ecosystems. In Chapter 6, the ecology of the common subclimax of grassland is discussed and that of the establishment of organisms in ecological niches. These lead to consideration of succession within an ecosystem.

Finally, acceptance of man as a species exercising a more profound influence on the surroundings than any other organism, leads to consideration of the ways in which he pollutes (i.e. degrades) the environment by the pesticides and other toxins he releases into it and how, by his introduction of organisms, that environment can further be changed.

1 The organism and its environment

A woodlouse, *Oniscus asellus*

1.1 Problems imposed by evolution

Woodlice have close relatives in the marine isopods. Although all are more or less terrestrial, their immediate ancestors were marine. They are a particularly interesting group in that the common species exhibit varying degrees of adaptation to conditions on dry land.

For small invertebrates, life on land has entailed the solution of a number of problems. No longer immersed in water from which oxygen can be extracted by some kind of gill mechanism and into which excretory products can freely diffuse, special organs of respiration and excretion must have been developed and, at the same time, some means for water-conservation evolved. The osmo-regulatory systems of the marine isopods allow a nice degree of balance between the internal and external media. In the case of terrestrial isopods, loss of water to the atmosphere can be disastrous.

Many land invertebrates have reduced water-loss by evolving an integument impervious to water-vapour, but this can induce other problems by inhibiting diffusion of oxygen and carbon dioxide. It also provides supports, but restricts growth, except by moulting.

The nature of the integument is a factor largely decisive in the ecology of the terrestrial arthropods, which are divisible into two groups, those with an impervious body-covering with its attendant drawbacks and those without. Woodlice lack an epicuticular layer of wax and belong to the second group. They lose water rapidly in dry air, and the degree varies from species to species.

Woodlice have adopted behavioural responses to stimuli in their surroundings which cause them to seek humid conditions and to lead a cryptozoic mode of life.

1.2 Identification

Various species of woodlice are common in gardens, on waste land, on farms and even at the margin of the sea. The last situation is colonized by the largest species, *Ligia oceanica*, which attains a length of 30 mm.

Ligia inhabits rock crevices, and although it will make nocturnal forays down the beach at low tide, by day it never wanders far from the high tide mark. It is, nevertheless, truly terrestrial and cannot survive prolonged submersion in the sea unless the water is well aerated. Belonging to the same family is *Ligidium hypnorum* which is smaller and is found inland but nearly always in the vicinity of water.

Woodlice belonging to the family Trichoniscidae are also to be found in very moist places while the Porcellionidae and Armadillidiidae are found in progressively drier situations.

Many ecological investigations need not involve time and labour in identification down to species level. However, where one is studying the autecology of a species, involving the relationship of that species with its environment, it is clearly necessary to be sure of the identity of the species with which one is dealing. A key to the different genera of British woodlice is given at the end of this chapter, *Oniscus asellus* being the only member of its genus.

1.3 Oniscus as a living organism

Structure

Before the effects of various environmental factors on the physiology and behaviour of an organism can be assessed, some knowledge of its structure is needed and particularly of those parts of the body which are significant in adapting the animal to its environment. Reference to Figs. 1.1 to 1.6 will help with the names given to the various parts of the body.

Oniscus is oval and flattened dorso-ventrally, the best body shape

---

Fig. 1.1 *Oniscus asellus:* I Male. 15 mm when fully grown. A drawing to show the body structures.

Oniscus asellus: II Male genitalia. (a) Endopodite of 1st pleopod of the right side. (b) Endopodite of 2nd pleopod of the right side. During copulation the projections (stylets) of the 1st and 2nd pleopods of the right side in the male are introduced into the left hand genital opening of the female. The male then shifts to the other side of the female and the procedure is repeated with the left hand stylets being introduced into the female's right hand genital opening. Since the whole of the mating procedure takes place in total darkness, it is very difficult to observe.

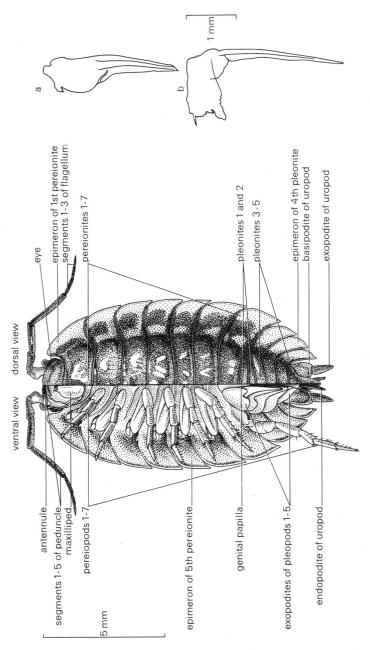

1 mm

a

b

dorsal view

ventral view

eye
epimeron of 1st pereionite
segments 1-3 of flagellum
pereionites 1-7

pleonites 1 and 2
pleonites 3-5
epimeron of 4th pleonite
basipodite of uropod
exopodite of uropod

antennule
segments 1-5 of peduncle
maxilliped
pereiopods 1-7

epimeron of 5th pereionite

genital papilla

exopodites of pleopods 1-5

endopodite of uropod

5 mm

3

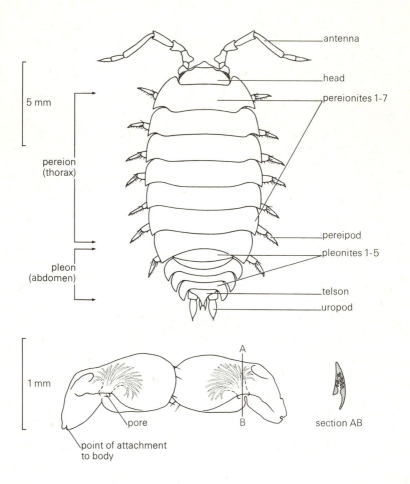

Fig. 1.2 *Porcellio scaber:* I Dorsal view of male. 15 mm when fully grown. A drawing to show the body divided into pereion and pleon.

Porcellio scaber: II Exopodites of the 1st pair of pleopods in the female to show the pseudotracheae. These are a mass of minute air-channels present in the more terrestrial species in which their function would appear to be the cutting down of water loss associated with the uptake of oxygen. The pseudotracheae appear as white patches in living specimens on the ventral surface of the pleon. Compare this drawing with the photograph (Fig. 1.6) in which the pseuodotracheae can be discerned.

Fig. 1.3 *Oniscus asellus.* Dorsal view of male. Note the three-jointed flagellum, the very pointed telson and the non-tuberculate integument.

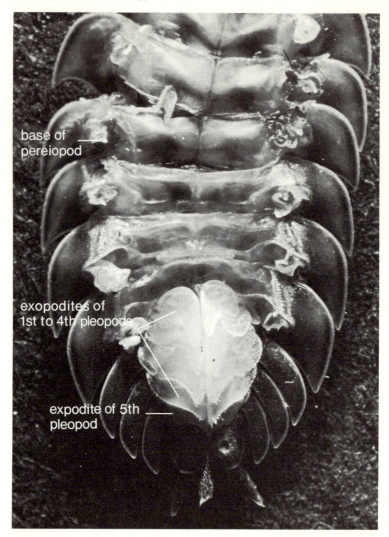

Fig. 1.4 *Oniscus asellus*. Male. Ventral view with pereiopods removed. The photograph shows the exopodites of the first four pairs of pleopods, the fifth pair being just distinguishable. Neither the endopodites nor the genital papilla, shown in Fig. 1.1, are visible.

Fig. 1.5 *Porcellio scaber.* Dorsal view of male. Note the two-jointed flagellum and the tuberculate integument.

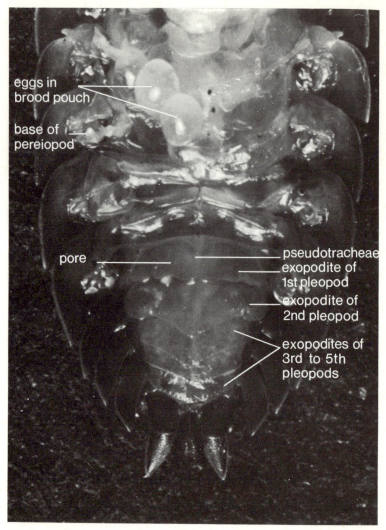

Fig. 1.6 *Porcellio scaber*. Female. Ventral view with pereiopods removed. The photograph shows the exopodites of the five pairs of pleopods. The first two pairs are enlarged and the pseudotracheae, which ramify the tissue of the exopodites, are visible. Also, the pore opening into the tracheae, can just be made out on the right hand exopodite of the 1st pair of pleopods. The others are not so easy to distinguish.

for a cryptozoan which spends most of its life in narrow crevices beneath stones, under logs or in the interstices of damp rubbish heaps. The head bears a pair of compound eyes and two pairs of antennae. The first pair is long, the flagellum having three segments (cf *Porcellio*). The second pair are the minute antennules not visible from the dorsal side and composed of three segments. The thorax, or pereion, consists of seven segments, each carrying a pair of legs. Pairs of ventral plates on segments two to five in the mature female form a brood pouch in which the eggs and newly hatched young are carried.

The abdomen, or pleon, is narrower than the thorax and is composed of six segments, each, except the last, bearing a pair of pleopods. The exopodites of the pleopods are greatly expanded, forming the chief surfaces from which water can evaporate. In some species, e.g. *Porcellio* (Fig. 1.2), this is the position of the pseudotracheae, tufts of invaginated tubules which can be seen as white patches.

In the male the first two pairs of pleopods are modified as copulatory organs, their inner rami being long and pointed.

The appendages of the sixth and last abdominal segment are called uropods and protrude posteriorally.

Life cycle
In common with other isopods, the female *Oniscus* carries its eggs within a thoracic brood pouch (marsupium) formed by pairs of plates arising from the second to fifth thoracic appendages.

The life cycles of several species of woodlouse, including *Ligia oceanica* and *Armadillium vulgare*, have been studied in greater detail than that of *Oniscus*. The number of eggs varies from seven per brood in *Trichoniscus* to one or two hundred in *Armadillium*. They are incubated in the brood pouch and the newly-hatched young resemble the parents except that they are paler and have six instead of seven pairs of legs. The appendages are tightly compressed under the body for the first few days, making movement impossible. At this stage they absorb moisture over their general surface from the brood pouch, the fluid in the pouch decreasing as their size increases. Heeley[35] found that the duration of the embryonic and larval stages are more or less constant for each particular species. In *Philoscia muscorum* the embryonic period is 21 days and the larval period 3 days while, for *Porcellio dilatatus*, the corresponding figures are 64 days and 9 days. The larval period of different species appears to increase approximately in proportion to the size of the brood.

In *Oniscus* the emergence of the young from the brood pouch may take 2 or 3 days and they are released through an opening in the pouch. At first the young tend to remain together and the female remains quiescent, but towards the end of the period of emergence the female becomes more active and forces out, by her body movements, any young still remaining in the marsupium. It is often possible to witness a sudden expulsion of the young from the pouch when a gravid female is handled, for she can accomplish this, if disturbed, by forcing the plates of the marsupium apart during rhythmic contractions of the body.

The first moult occurs after the first 24 hours of freedom and the young, now with seven pairs of legs, enter a critical period of their lives, for if the environment is too dry at this time, they quickly die through desiccation and if too wet, fungal diseases can attack them.

The longevity of woodlice varies with the species, but most do not breed until their second year. *L. oceanica* lives for about 3 or 4 years and during its breeding life can produce at least five broods.

Moulting takes place in two stages, the exuvia of the anterior part of the body is cast some time after that of the posterior part. Despite presenting an odd appearance. The two-stage moult confers a degree of protection against predators.

Predation and control of numbers

Oniscus is eaten by birds, lizards, slow worms and toads. Among its invertebrate enemies must be numbered spiders, harvestmen, centipedes and mites. The spiders *Dysdera crocota* and *D. erythrina* have strong jaws capable of crushing the cuticle which, because of its chitinous content, proves too tough for many invertebrate predators. In addition, acid-smelling secretions of the tegumental glands act as a deterrent, apparently making the animal distasteful, especially to many species of spider.

Parasites, and in particular dipterous parasites, do not seem to play an important role in the control of woodlice. But there is evidence that cannibalism occurs in over-crowded situations and may be a significant factor in controlling populations.

1.4 Behaviour under natural conditions

Invertebrates often respond to stimuli without actually moving in the direction of the stimulus, which merely affects the activity of the animal. Such a response is a *kinesis*. *Orthokinesis* is a response resulting in alteration of speed of movement while *klinokinesis* is one in which the animal turns or changes direction while it is moving. It is important to distinguish between kineses and *taxes*, the latter being responses made either towards or away from the origin of the stimulus.

Since there is a lack of waterproof integument *Oniscus*, in common with other woodlice, is extremely sensitive to humidity gradients, and tends to congregate in areas of high humidity. This manifests itself in a decrease of speed in moist conditions but once they have located a damper situation, individuals are inclined to remain there. However, the relative humidity of the surrounding air and their negative phototaxis causes them to congregate together, seeking out narrow crevices in wood or other damp places and often making close contact with each other and with their immediate surroundings, thereby avoiding excessive evaporation.

Oniscus tends to seek out the darker, damper regions by day, venturing out at night, when dispersal to new environments can take place. The extent of nocturnal wandering is dependent upon ability to withstand water loss. Excursions during darkness to drier places is accompanied by an increase in photo–negative behaviour so that once daylight comes, a return to cover is made.

Should the surrounding become drier during daytime, *Oniscus* can adopt a positive phototaxis causing it to wander around until it finds a damper place in the dark. Once there, it again becomes negatively phototactic.

1.5 Water loss

It has already been mentioned that loss of water is an important factor in the survival of terrestrial isopods. It can take place in several ways.

Since the integument lacks a waterproof covering, water can be lost directly through the body covering and the rate at which this occurs will be affected by (1) the size of the animal, (2) its behaviour and (3) the saturation deficit of the surrounding medium. Obviously evaporation takes place more rapidly from a larger surface than from a smaller. By rolling into a ball, *Armadillium* and other species capable of doing this, reduce their exposed surfaces and thereby cut down the rate of evaporation. Bunching of woodlice in groups also has the same result. The effect of the saturation deficit of the air, however, is more complex.

The degree of dampness of the air is often called *relative humidity* and can be defined as the ratio of the mass of water vapour per unit volume of air to that of a similar volume of saturated air at the same temperature. It is usually expressed as a percentage. The *saturation deficit*, or the actual drying power of the air, is the amount (measured in g m^{-3}) by which the water content of the air is lower than that of saturated air at the same temperature and pressure and can be determined from readings on a wet– and dry–bulb thermometer. From the biological standpoint, this is a better measure of humidity because it does not fluctuate during the day with changes of temperature.

Measurement of the actual rate of evaporation from a surface in a small space is difficult: it is affected by such differentials as the nature of the surface, wind, temperature, and the saturation deficit. Furthermore, such physical conditions are not necessarily decisive in the survival of organisms living there.

The fact remains that, in *Oniscus*, an important factor in survival is its ability to reduce evaporation, either by staying in an atmosphere of near-saturation or by taking 'avoiding action'. Moulting is a time of special risk: generally in arthropods it coincides with inactivity when predation might be expected to account for much loss, but hazards are partly mitigated by the isopod habit of shedding the exoskeleton in two halves. Measurements by Webb-Fowler (unpublished) have shown a slight increase in water-loss after both the posterior and anterior halves have been cast, but it appears to be unimportant.

Excretion and defaecation
Segmental glands of the second maxillae are the chief excretory organs in *Oniscus*. The loss of products of nitrogen metabolism might be expected to be accompanied by the loss of water, but evidence for this is slight. Faecal pellets deposited after food with a high water content has been taken, are often particularly moist. But as far as is known, there are no structures comparable with the rectal glands of insects, which have a reabsorptive function.

Respiration
The pleopods are important organs for the absorption of oxygen

and loss of carbon dioxide by diffusion. Rapid loss of moisture from their surfaces impedes gaseous exchange. It has been shown that other parts of the body are probably capable of oxygen absorption.

1.6 Water gain

When foraging, *Oniscus* is at particular risk, and it would seem to have means of acquiring water even when its movements take it out of a saturated atmosphere. Obviously imbibition of water when moist food is available is one. It can survive indefinitely in unsaturated air provided it can make contact with a moist surface such as a block of plaster of Paris or a slice of potato. Evidently it can take in liquid by 'drinking'. There is also absorption of water through the anus. An additional method is by capillarity, and it is likely that this is the means by which eggs in the brood pouch are kept moist.

1.7 Temperature

Migration to land has presented woodlice with greater temperature ranges, and faster temperature changes, than those encountered by their aquatic counterparts. Animals dwelling in soil avoid low temperatures at the surface by moving down and reaching levels where the ground is warmer. (Periodically *Oniscus* frequents certain kinds of leaf litter and thereby merits comparison as a migrant). Little work has so far been done in field conditions on low temperatures which are lethal for woodlice, although upper lethal temperatures have been determined. In saturated air, *Oniscus* can survive for 24 hours at 31.5 °C. Exposure for short periods to unsaturated air actually increases the animal's tolerance of high temperatures because of the cooling effect of the evaporation of body water: Edney[20] found the depression of the body temperature of *Oniscus* to be 0.4 °C when it was exposed to dry air at 20 °C.

Under laboratory conditions, experimental work may give misleading results by over-emphasising the effect of a particular factor which, in natural conditions, may be less significant. In the controlled situation of a laboratory, the rate at which water is evaporated seems to be important in determining the temperature of the animal's body. In the field conduction, convection and heat radiation complicate the effects of water loss: their interaction probably accounts for the survival of *Oniscus* during its short

13

sorties into warm sunlight.

Such findings emphasise the need for investigations to be undertaken in the first place under natural conditions. Thereafter certain environmental effects can be tested in the laboratory with the intention of verifying, under controlled conditions, specific reactions to separate stimuli.

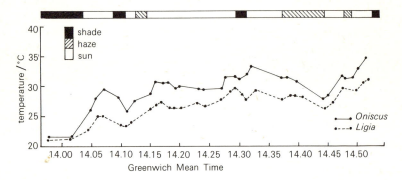

Fig. 1.7 THE INTERNAL TEMPERATURE OF *Oniscus asellus* AND *Ligia oceanica*. The woodlice were exposed simultaneously upon a wooden substrate, to varying degrees of light. (Air temperature 5 mm above substrate followed closely that of *Oniscus*. R.H. 55–63 percent; wind speed c. 50–100 cm/sec.) After Edney[20].

Figure 1.7 (after Edney[20]) shows the internal temperature of *Oniscus asellus* relative to the air temperature 5 mm above a wooden substrate. These figures were derived by exposing dried specimens alongside living specimens on a substrate which was moved from sun to shade. The temperature of the dried specimen was found to be always higher than that of the living specimen and the degree of depression in the latter corresponded approximately with the rate of water loss.

1.8 Distribution

The fact that *Oniscus* shows negative photokinesis and a positive orientation to humidity might result in the animal being trapped in damp, dark crevices from which it could not emerge to make feeding forays. *Oniscus* appears, however, to be little affected at low light intensities and there is probably a reversal of phototaxis in dry conditions. Cloudsley-Thompson[10] has shown a positive hygrokinetic response in this species after a rise in temperature in strong light, which may account for the fact that *Oniscus* spends much of the day in dark, damp retreats, becoming active by night when dispersal takes place.

The results of these complex reactions to humidity, temperature and light, all play their part in the distribution of *Oniscus*, and in winter may induce lateral migration to compost heaps, cellars, greenhouses and so on. Whatever the movements, whether diurnal or seasonal, the animals are influenced by the environmental factors to which they have most recently been exposed, and this must be taken into account in designing either field or laboratory tests.

1.9 Experimental work

The foregoing account of the structure, physiology and behavioural responses of *Oniscus* must form the background information, but will not necessarily provide the answers to experimental work carried out to test the survival value of such responses.

Oniscus has been selected as an organism for study, not only for its ubiquity but because it presents behavioural problems which cannot always be answered in terms of a reaction to a single habitat factor. Indeed this is never the case. Since in the field it may react very differently from its behaviour in the laboratory, investigations in the field can only suggest a line of approach or promote hypotheses to explain reactions.

Field observations
Oniscus asellus is usually to be found in quite large numbers under heaps of rotting wood or among a pile of stones or bricks. In colder weather there are often fairly large colonies in greenhouses.

Questions will at once arise as in what kind of situation *does Oniscus* live and how large is the population? Are they congregating by day in the obviously damper parts of the area? Are they active or more or less immobile? Are they to be found in twos and threes or bunched together? Are there any predators to be seen such as spiders, centipedes, toads or even hedgehogs? Are there young woodlice among the colony? The answers to these general observations will suggest more detailed investigations such as the effect of humidity gradients or reactions to temperature changes, and so forth.

One Approach to Ecology

Estimating populations

Estimation of the size of a population of woodlice can be made by the method of capture–recapture or the *Lincoln Index* method also described in Chapter 8. This involves the collection of a random sample of a large number of individuals (say 50) from the population. Marking is best done by using a bright coloured waterproof paint to make a spot on the dorsal surface of the woodlouse. The animals are then returned to the colony. After time has been allowed for them to mix with the rest of the population, a second sample is taken and the number of marked individuals in this sample is noted. Assuming that the marked individuals have mixed at random with the rest of the population, the size of the total population (p) can be estimated as follows: $p = n \times \frac{s}{r}$ where $n =$ number in the first sample, $s =$ number in the second sample and $r =$ number of marked individuals in s. This technique also assumes that no woodlice have emigrated, immigrated or multiplied.

Controlled environments

In a cryptozoic animal such as a woodlouse, it is probable that one

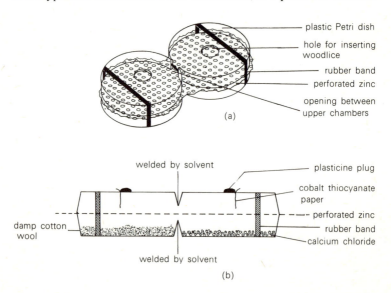

Fig. 1.8 A CHOICE CHAMBER FOR STUDYING BEHAVIOURAL RE-SPONSES OF WOODLICE TO DIFFERENT STIMULI. (a) Choice chamber constructed from two plastic Petri dishes. (b) Section through choice chamber.

16

of the most important factors influencing its distribution will be humidity and this assumption will have been borne out by the investigations made under natural conditions.

Tests of the behaviour of *Oniscus* under controlled conditions can be made by using choice chambers which provide the animal with two alternatives–for instance wet or dry, light or dark situations.

The type of choice chamber illustrated in Fig. 1.8 can be constructed from the bases of 4 Petri dishes. A small hole is cut in the sides of two of them and also a central hole in each. These form the two upper chambers. These two halves are then welded together using a perspex solvent (ethylene dichloride). The other two Petri dishes are also welded as shown in the diagram. A sheet of perforated zinc is placed over the lower sections and the upper sections are held in place over the zinc by means of two stout rubber bands. Similar choice chambers are described by Waterhouse[82].

Reaction to humidity
Anhydrous $CaCl_2$ is placed in one of the lower compartments and cotton wool soaked in water in the other.

Woodlice of the same species and size must be exposed to the same humidity for several days before the start of the experiment, in a desiccator over an appropriate humidity solution. Five of these woodlice are then introduced to each compartment and the holes plugged with Plasticine which also holds a cobalt thiocyanate paper to check the humidities. Under normal conditions, humidities will be relative and not absolute. After 15 minutes the number of woodlice in each compartment is noted. The experiment should be repeated several times using fresh woodlice.

Reaction to light
The same choice chambers can be used to test reaction to light and dark by painting one chamber with Japlac enamel.

The graph (Fig. 1.9) shows the average result of actual experiments exposing full–grown specimens of *O. asellus* and *P. scaber* to dry and damp choice chambers and the graph (Fig. 1.10) the average of results obtained when exposing the two species to light and dark choice chambers. Both species responded more positively to illumination than to humidity but showed that *P. scaber* was slower to settle down than *O. asellus* and was less critical in its choice of conditions. These results might well tally with observations

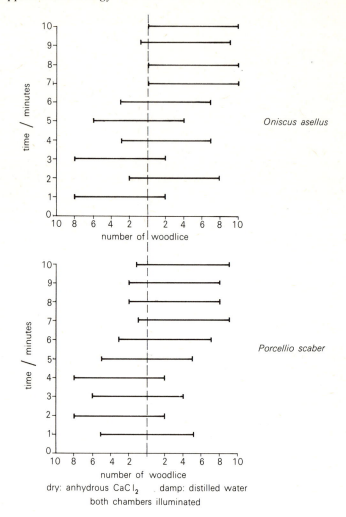

Fig. 1.9 COMPARATIVE BEHAVIOUR OF *O. asellus* AND *P. scaber* IN RE-SPONSE TO HUMIDITY. The broken line down the centre indicates the neutral position, i.e. that of the actual passage-way between the two chambers. The tendency for *P. scaber* to be less clearly defined in its 'choice' may correlate with its more arboreal habits.

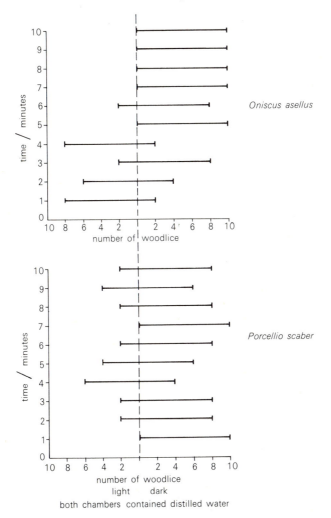

Fig. 1.10 COMPARATIVE BEHAVIOUR OF *O. asellus* AND *P. scaber* IN RESPONSE TO LIGHT AND DARK. The broken line down the centre indicates the neutral position, i.e. that of the actual passage-way between the two chambers.

of their natural distribution – *P. scaber* often being seen in exposed situations such as on the dry branches of trees during the hours of daylight, usually shunned by *O. asellus*.

19

Reaction to more than one stimulus

By using solutions to give different humidities (see Richardson[64])
in each side of a series of chambers and at the same time subjecting
each chamber to different light intensities, an estimation can be
made of the reaction of the animals to the two stimuli and an idea
can be gained of the hierarchy of the stimuli. If results are to be
analysed then the index chi squared (χ^2) must be used. (See Bishop[5]).

The choice chambers can also be used to test food preferences.

Orientation

Reference has been made in Section 1.4 to the fact that woodlice
show both kinetic and tactic behaviour. To investigate these patterns
of behaviour a larger chamber than that used in the previously
described investigations is really needed. A uniform environment is
provided by placing similar solutions on either side of the chamber.
Comparisons of the kind and speed of movement in different humidities are made by marking movements of the woodlice with a
chinagraph pencil on the lid of the chamber, at the same time using
a stop watch. Movements and behaviour are, however, affected by
the walls of the chamber, thigmotaxis often interfering with kinesis.

After trying out these various behavioural responses it will be
evident that unless very exact instruments of measurement and more
complicated apparatus is used, they can only give an idea of the
nature and degree of the reactions to different stimuli. Nonetheless
they will have served one important function by demonstrating how
difficult it is to make accurate measurement of the responses and of
the microclimate. But at the least they will show the levels of
importance of the stimuli as well as the value or negation of statistics
when applied under these circumstances.

Some conclusions

The general picture that emerges is of an animal ill-equipped for a
terrestrial existence. The lack of a specialized structural protection
against evaporation from the body surface, combined with relatively
inefficient organs of respiration, which are largely dependent for
functioning upon a saturated or nearly saturated atmosphere, must
condemn *Oniscus* and its relatives to a life devoid of opportunities
for moving far beyond the confines of a restricted environment.
Deficiences also in its mode of reproduction likewise contribute to
its limited life, for the developing eggs and young must be kept moist

within the brood pouch.

Despite these apparent disadvantages, large colonies of woodlice are often to be found and during the late summer there may be vast numbers of young individuals feeding and growing in the moister regions of compost heaps. We must look for the answer, not so much to their morphological or physiological adaptations, but to their behavioural responses which restrict them to an ecological niche where the hazards of a true terrestrial existence are reduced.

The investigations suggested using choice chambers will have shown that woodlice 'prefer' damper to drier conditions. But what is the mechanism by which, in nature, they find such conditions and where are the organs of perception located? Little is known about the way in which woodlice are made aware of their surroundings and the part played by the antennae and eyes in sensory perception.

It might be said that terrestrial woodlice form one of the numerous links between sea and land. Although the geological age of the group is uncertain, there were probably terrestrial oniscoids even before the Tertiary period. Lack of apparent progress in their evolution since those times can possibly be attributed to the retention of a water-permeable covering which is essential for respiration and as a means of cooling when exposed briefly to high temperatures.

See also Cloudsley-Thompson[10], Cloudsley-Thompson and Sankey[12], Kevan[39] and Sutton[76].

1.10 Key to the common genera of British woodlice

(The distinguishing features of each genus are shown in Fig. 1.11)

1 Flagellum of antenna with ten or more segments   2
   Flagellum with less than ten segments   3
2 Outline of pereion and pleon continuous. Length of animal
   up to 30 mm. Coastal, near high water spring level of shore
   but wanders onto foreshore on the ebb tide at night   *Ligia*
   Pereion broader than pleon, making a distinct break in
   the outline. Smaller animals 6–10 mm long. Restricted
   to very damp moss in woods   *Ligidium*

3 Flagellum with 4 segments   4
   Flagellum with less than 4 segments   5

types of flagellae (shaded)

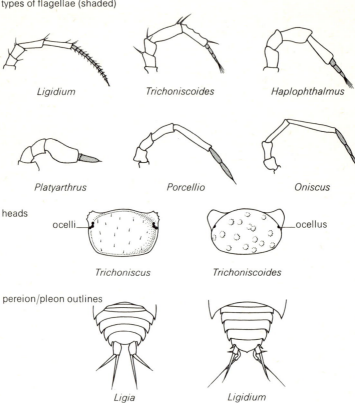

Ligidium

Trichoniscoides

Haplophthalmus

Platyarthrus

Porcellio

Oniscus

heads

ocelli —

— ocellus

Trichoniscus

Trichoniscoides

pereion/pleon outlines

Ligia

Ligidium

shapes of telsons and uropods

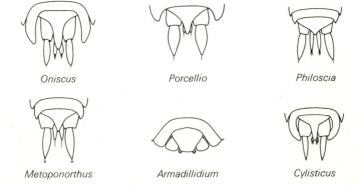

Oniscus

Porcellio

Philoscia

Metoponorthus

Armadillidium

Cylisticus

4   Single ocellus on each side of head, poorly developed and
    often invisible. Body 3–5 mm. long. Ping or white. In
    moist woodlands under moss, bark or fallen leaves
                                                    Trichoniscoides

    3 ocelli on each side of head distinguishable under a lens.
    Body reddish brown or whitish with brown markings. Very
    moist places among roots of plants. 2–4 mm long.
                                                    Trichoniscus

5   Flagellum with 3 segments   6
    Flagellum with 2 segments   8

6   Eyes each of 1 ocellus. Lateral lobes of head well developed.
    Raised ridges composed of tubercles on pereion. Uncommon.
    Up to 5 mm long.   *Haplophthalmus*

    Eyes compound   7

7   Lateral lobes on head absent. Body up to 10 mm. Smooth,
    shiny integument. Telson not produced to a point. In
    rubbish heaps or under dead leaves. Up to 10 mm long.
                                                    Philoscia
    Lateral lobes on head strongly pronounced. Body 15 mm.
    Smooth. Telson produced to a point. Pleopods without
    pseudotracheae. Ubiquitous. Found in moister situations
    than *Porcellio*.   *Oniscus*

8   Eyes absent altogether. Body colour white. Up to 3.6 mm.
    Epimera well developed. Usually found in ants nests.
    Often common.   *Platyarthrus*
    Eyes compound.   9

9   Uropods short and not projecting beyond telson, their
    endopodites expanded laterally forming with the telson a
    convex covering posteriorly. Outline of pleon and pereion
    continuous. Body strongly convex. Capable of rolling into

Fig. 1.11 SOME FEATURES BY WHICH THE DIFFERENT GENERA OF
WOODLICE CAN BE DISTINGUISHED.

a ball. Widely distributed. Abundant on calcareous soils.
Up to 20 mm long. Uropods project beyond telson.

Armadillidium

Uropods project beyond telson.   10

10 Telson produced into a point   11
Telson triangular, never produced into a point. Body up
to 9 mm long, covered with fine bluish-grey bloom which
becomes dark brown when rubbed off. Widely
distributed and fairly common   *Metoponorthus*

11 Pseudotracheae on all 5 pairs of pleopods. Endopodites of
uropods not expanded laterally. Capable of rolling into a
ball. Slaty grey with irregular light markings. Fairly
common. Associated with calcareous soils. Body 15 mm long.

Cylisticus

Pseudotracheae on first two pairs of pleopods. Not capable
of rolling into a ball. Integument tuberculate. In compost
heaps, usually in the drier parts. Very common. Body up to
17 mm long.   *Porcellio*

2 The organism and its environment

Creeping buttercup, *Ranunculus repens*

2.1 Distribution and morphology

Assuming the synonyms buttercup and crowfoot to apply to all members of the genus *Ranunculus* whose flowers are in inflorescences and have yellow petals in a single whorl with a recognisable calyx, creeping buttercup = creeping crowfoot (*R. repens*), meadow crowfoot (*R. acris*), bulbous crowfoot (*R. bulbosus*) and corn crowfoot (*R. arvensis*), together comprising what are generally accepted as the common buttercups of fields and meadows in Britain. The first three are perennials and are widely distributed in permanent grassland: *R. repens* and *R. acris* alike have a regional distribution which is practically universal; *R. bulbosus* is predominantly an English buttercup, commonest south of the Scottish border and east of Wales and the Dartmoor dome; and *R. arvensis* is a buttercup of southern (lowland) Britain where it has an interrupted distribution correlated with its frequency as a cornfield weed on calcareous soils (Perring & Walters[58]). Like the crops with which it is particularly associated, *R. arvensis* is an annual. It is decreasing.

A key to the yellow-flowered buttercups appears at the end of the chapter.

Harper[33] has summarised the status of *R. repens* in our flora. Although commonly regarded as the buttercup *par excellence* of such wet situations as low-lying meadows on heavy soils, it also occurs in damp woods, in cornfields having a chalky substrate, on dune slacks and gravel heaps, and is frequently a troublesome horticultural weed, notably in damp and somewhat shaded parts of flower-beds and kitchen gardens. In view of such an environmental range, its structural uniformity is remarkable. Partly on this account, references to it in standard floras tend to be brief. In contrast, *R. acris*, the species with which it seems to have closest relationship and which shares in some degree its apparent affinity for moist conditions, is markedly polymorphic, having a number of named varieties and subspecies. Sowerby & Sowerby[74] recognise three main variants and Clapham, Tutin and Warburg[9] list four. Nevertheless, visible differences, which are more or less constant, appear between those plants of *R. repens* growing in open situations and those in moist and shady places. That the forms are phenotypes is suggested by experiments we have carried out; as far as we can discover, Sowerby was the first to draw attention to them. In Clapham, Tutin and Warburg, the chromosome numbers are given as $2n = 16, 32$.

Features diagnostic of creeping buttercup include a combination of the following (Fig. 2.1). Initially, in succession to the primary root, the plant produces a cluster of up to 25 adventitious roots from the base of a short, stout stock. Thereafter, strong epigeal stolons (runners) radiate from the axils of the radical foliage and root at their nodes: lateral stolons also develop from the axillary buds of some of these nodes: the result is a meshwork of superficial stolons lying on the ground and often concealed in the surrounding herbage. The flowering stems are erect, up to 60 cm tall, and carry flowers 2–3 cm in diameter with 5 bright yellow glossy petals and 5 sepals spread out in contact with the underside of the corolla, and terminating the branches of an irregular cymose inflorescence on furrowed stalks. Compared with the inflorescences of *R. acris* and *R. bulbosus*, the flowers tend to be sparse. In outline, the basal leaves and those of the lower stem are broadly triangular and consist of three leaflets of which the terminal one has a longer stalk than the other two and projects well beyond them: the leaflets are further divided into three-toothed segments: the upper leaves are sessile with narrow, entire segments. Other than the petals and achenes, all the parts above ground have loose spreading hairs.

Normally the first flowers appear in May and inflorescences persist until late autumn: an area of *R. repens* in Worcestershire observed over a period of seven years has produced some flowers continuously in mild seasons from April to the end of January.

Where the atmospheric conditions are those of high relative humidity, the substrate is wet and the illumination low, forms may

Fig. 2.1 CREEPING BUTTERCUP (*Ranunculus repens*). Entire plant. Its older leaves retain the fundamental triangular form even when their leaflets become complex.

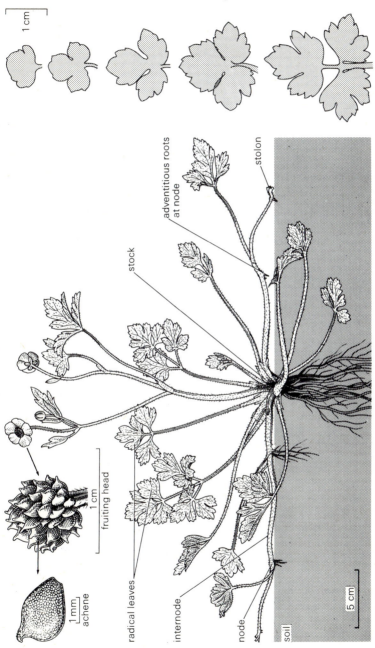

develop which exhibit the variants noted by Sowerby. These include three features particularly:

1 stems are less hairy and, in extreme cases, are glabrous;
2 runners ascend instead of lying prostrate and do not produce roots at the nodes;
3 aerial parts in general are larger and apparently more robust.

Sowerby's imprecise observations need qualifying.

Seeds from plants showing some of these variants and growing in damp woodland on boulder clay, gave rise to normal vegetative forms when sown in open ground on Kuyper marl, where illumination was high, and the converse also held good. The only justification found for Sowerby's claim that the runners' nodes are unrooted was that, in deep shade, the runners became comparatively long and rooting developed at the distal before the proximal nodes. It is interesting to compare this with what has been found by other workers for certain woodland species, including bugle, creeping jenny and yellow archangel, where moist, shaded habitats favour vegetative spread by runners and drier, illuminated situations promote the development of flowers. Salisbury's findings in this connection on yellow archangel are considered later (page 44).

Etiolation, and/or luxuriance resulting from shelter and abundance of moisture, could be responsible for some of the effects, notably the ascending form of the runners and the relative bushiness of the foliage. Reduction in hairs in damp conditions is interesting in view of the suggestion (Darlington[18]) that hairiness might be taken as a *prima facie* indicator of xerophytism.

Further work is needed on the relationship between stolon-development and seed-yield in *R. repens*. A comparison between the relative importance of vegetative reproduction and seeding of a species makes a worth-while project, giving insight into the nature of ecological aggression (see Ashby[3]), as also into the physiological balance between the vegetative and the reproductive states.

2.2 Reproduction by flowering

Clapham, Tutin & Warburg describe flowers of *R. acris* as more or less homogamous, with some forms self-sterile and others gynomonoecious; features which might apply also to *R. repens*. So little work seems to have been attempted on the processes of pollination and

fertilization specifically in *R. repens* that the entire subject needs
experimental investigation. Both species are visited by a range of
insects, notably solitary bees and the smaller hoverflies (syrphids).
Bees from Worcestershire samples of *R. repens* have included the
miners *Andrena timmerana*, *A. cingulata* and *Panurgus calcaratus*;
an addition from maritime counties only has been *Dasypoda
hirtipes*. The last two, *Panurgus* and *Dasypoda*, are noteworthy for
two reasons: both normally show preference for composite inflorescences; and from their behaviour might be expected to be
functional pollinators of many of the flowers they reach, since they
cause great disturbance to the essential parts by bathing themselves
in the pollen and executing virtual somersaults as they collect it with
the whole body. Common syrphids on the Worcestershire material
have included *Chilosa albitarsis* and *Criorrhina ranunculi*: *Helophi-
lus pendulus*, 14 mm long, was the largest.

It has been suggested that, by comparison with other *Ranunculus*
spp, fruit yield in *R. repens* is sparse as a result of its vigorous
vegetative production of runners. Species that multiply vegetatively.
for example dog's mercury, commonly set few viable seeds. Creeping
buttercup has fewer flowers per inflorescence than the other grassland buttercups, although this could partly be offset by its longer
period of flowering. Salisbury[67] gives the number of achenes
commonly developing from an individual flower as 25–30 in *R.
repens*, 20–30 in *R. bulbosus* and 30 in *R. acris*, and mentions one
large plant of *R. acris* which produced 22 000 fruits in a single season.
We have no corresponding total for *R. repens*: in a species giving
rise vegetatively to a number of inflorescence-bearing units it would
be difficult to determine the limits of single plants. Counts made
from fruiting heads of *R. repens* gathered during July-August in the
west Midlands suggest a value slightly lower than Salisbury's: 100
heads were taken from each of three situations: a) garden soil in
open and well-drained places; b) heaps of small stones set up by
roadsides for icy conditions but undisturbed for several years: and
c) the damp and sunless base of an old wall, accumulating run-off
from precipitation, where the phenotypes were established which
have already been noted. Averages were obtained of 26 for a, 21 for b
and 31 for c, and an overall mean for 300 heads of 26. When broken
down, the figures point to a survival potential in some of the
differences: thus c, where there were some stolons which were non-
rooting and dispersal by asexual means apparently reduced thereby,

gave the highest number of sexually produced bodies despite the possibility of partial concealment from flying insects. The low value for b could reflect some kind of nutritional deficiency.

As they stand, the figures clearly have limited significance for several reasons. One derives from the fact that all the samples came from places liable to greater disturbance than the permanent grasslands of which *R. repens* is a frequent constituent. A second is that no account is taken of the actual germination potential of seeds formed in the various situations although, in arriving at the figures, any achenes which appeared undersized or malformed were discounted. A third objection could arise from the time of year when the counts were made, on the grounds that *R. repens* has a sexual phase persisting until late in the year and that samples taken after August might give different results. In fact, checks were made in all three places to meet this. Apparently the only difference was that fruiting heads became progressively harder to find from early October onwards, but that any which were examined gave the same results as before.

2.3 *Dispersal of sexual propagules*

Clapham, Tutin and Warburg note the length of a *R. repens* achene as 3 mm and Salisbury[66] as 3–4 mm, but we have found fruits as short as 2 mm. The seed is an oval body with a long axis of 2 mm: seeds in the smallest achenes are undersized and germination experiments have shown them to be non-viable. The only achenes we have seen with a length of 4 mm have come from the phenotypic variants with non-rooting runners: their thickness in the transverse plane is 1.0 mm, the same as for the 2 and 3 mm fruits; their depth 2.5 mm compared with an average of 2.0 mm for the medium-length achenes. Assuming Salisbury's figure of 0.0024 g is reliable for the average weight of a single achene in the 3–4 mm range, our findings of limits from 0.0019 to 0.0026 g for those of the 4 mm size only point again to a tendency in *R. repens* for dispersal of sexual propagules to be favoured when vegetative expansion is inhibited. The physiological changes which redirect the plant from vegetative into reproductive growth involve structural changes at the apex: presumably inhibition of stolon production represents a change in the hormone balance at the now reproductive apex.

Compared with the smaller achenes, the bigger are broad and flat and present a larger surface to air-flow, and their weight does not increase in proportion to their size. To test the efficiency of achene dispersal with particular reference to wind, a plot 15 × 5 m, was prepared of clayey soil, used previously for growing kitchen vegetables (Fig. 2.2). It had become weed-infested and creeping buttercup was abundant. The plot was surrounded by an admixture of grasses and *R. repens*: here buttercup was controlled by application in spring of MCPB 4 (2-methyl 4-chlorophenoxy)–butyric acid administered as 'Buterex'. The plot was subdivided into a row of three squares of equal sizes lying west-east (i.e. in the direction of the prevailing wind): the middle one was left undisturbed; the adjacent squares were turned over in spring of the first year by a rotivator and growths of all vascular plants appearing therein subsequently removed by hand-weeding; and a fence 2 m high and running north-south was erected between the central and eastern squares. This was divided into three equal sections, that in the middle consisting of a single strand of wire and those at the ends respectively of wire-netting with 2 cm mesh and of solid wooden panelling. In the first, second and third years, any runners invading the squares were cut

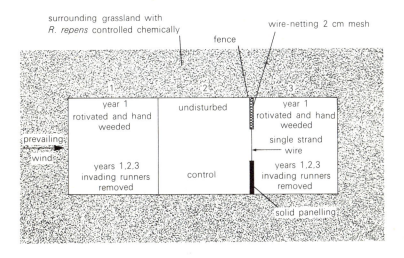

Fig. 2.2 DISPERSAL OF WIND-BORNE ACHENES. Experimental plot set up to test the dispersal of *R. repens* achenes in the field by wind. Dimensions of rectangle, 15 × 5 m. Malvern, Worcs.

off, and in the second and third years the positions recorded of all *R. repens* seedlings appearing in the exposed soil. From time to time, seedlings of other species were pulled up.

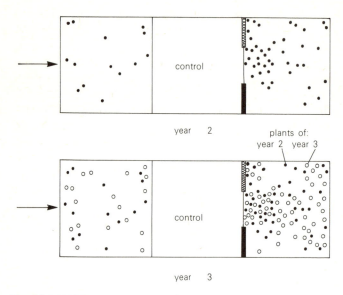

year   2

plants of:
year 2   year 3

control

year   3

Fig. 2.3 DISPERSAL OF WIND-BORNE ACHENES. Distribution of *R. repens* seedlings in second and third years. Assuming the parents to lie in the control section, the form of scatter confirms the expectation of low-level movement with wind a significant agent of transport. Malvern, Worcs.

Results are presented in Fig. 2.3. They suggest the efficacy of wind in bringing about dispersal, a difference being apparent between the windward and leeward sides of the fence. They also indicate the rather surprising fact that wire-netting of this particular mesh proved to be a significant wind-break. They do not eliminate the possibility of some of the growths developing from seeds with a dormancy period greater than two years, nor of dispersal by animals and other moving objects.

Viability and dormancy are facets of the same feature. Studies of the populations of viable seeds buried under diverse types of pasture and in various soils have been undertaken by several workers. From their investigations, Champness & Morris[8] came to the conclusion that viability is better preserved under acid or waterlogged conditions than where the substrate is basic, drier or better aerated: they determined that an acre of soil contained between five and six million seeds of creeping buttercup which are dormant but viable. Harper[32] recognizes three categories of dormancy, innate, induced and enforced.

Innate dormancy, a feature of the ripe seed as it leaves the parent, is genetically controlled. A seed having this dormancy cannot germinate until it has undergone after-ripening treatment or has received some specific stimulus. Exposure to light of the seeds of some buttercups may be relevant. Stimuli which break innate dormancy tend to be most active at the soil surface: alternate freezing and thawing as an agent in rupturing the testa, for example.

Induced and enforced dormancy depend on the interaction of the seed with its environment. *Induced* dormancy develops when a seed is exposed to certain conditions and continues after the conditions change, e.g. white mustard (*Sinapis alba*) exposed to high concentrations of carbon dioxide. Some special treatment is needed to break it. *Enforced* dormancy applies when a seed is prevented from germinating simply by environmental limitations, and as soon as these are removed – or the seed is removed from them by being brought to the surface during ploughing – it grows.

Further work to evaluate the categories of dormancy applicable to its seeds is needed on *R. repens*.

Animals as agents effecting dispersal include those which transport propagules on their surface: man and his vehicles can be considered here. Investigations by Darlington[18] demonstrated the importance of motor-tyres: washings from these taken in June yielded mud from which 13 angiospermous species were grown, the most numerous being *Poa annua* (387 plants), *Stellaria media* (274) and *Matricaria matricarioides* (220); and the same sample gave 84 seedlings of *R. repens*. The fruiting heads of *M. matricarioides* bear adhesive propagules which tend to stick to such solids as tyres and footwear: the achenes of *R. repens* have no such adaptations and, in contrast to the burr-like fruits of *R. arvensis*, also lack projections which might facilitate attachment to clothing or fur. In fact, they are particularly smooth and presumably had merely become caught up in the dirt sticking to the treads.

However, the turn-ups of trousers could be even more significant as a means of human epizoic transport. Darlington has investigated the living material collected in the trouser turn-ups of boys walking daily across fields to school; he spread the collected fluff over sterilized soil and incubated it at room temperature. Some 70 species were obtained in this way: during a single year (1938) the three highest totals were found to be grass spp 19 per cent, cleavers (*Galium aparine*) 14 per cent and *R. repens* 11 per cent. Two of these results are to be expected – the lightness and abundance of grass fruits in summer and the clinging hooklets on *Galium* – but the high value for *R. repens* is perhaps more surprising. With the exception of such adherent forms as *Galium*, many of the fruits and seeds are loose in the turn-ups so that the wearer becomes a sort of peripatetic censer mechanism for scattering propagules, notably the smoother kinds (*R. repens* and others), as he walks about.

All three grassland buttercups contain the glycoside ranunculin, which releases the unsaturated lactone, protoanemonin, when their tissues are crushed. Protoanemonin is a blistering agent, toxic to other vascular plants and antibiotic to bacteria and fungi. The sap of all three has been compared for protoanemonin content: *R. bulbosus* contains a high concentration, *R. acris* less and *R. repens* least. Shearer[68] reports that protoanemonin causes blistering of the lips and tongue of a grazing animal and also produces intestinal disorders. Both *R. bulbosus* and *R. acris* have proved fatal when eaten in quantity by cattle, but when dried and converted into hay they become innocuous. *R. repens* can be taken green in moderation without ill effects.

The dung of animals can be a significant vehicle for dispersal, as shown in Salisbury's table, which lists 51 species of weeds as having been recorded in the faeces of a limited range of farm animals. Doubtless many more plants have been missed. *R. repens* is the only buttercup noted (perhaps because of its low toxin content); Salisbury gives its seeds as having been germinated from manure of cattle and horses, but we can add that of goats and, perhaps rather surprisingly, of badgers from one locality (Malvern College Nature Reserve, Herefordshire). The extent to which passage through the digestive tract affects viability would be a profitable subject for an additional project.

Salisbury also lists the seeds of *R. repens* among the 32 kinds grown in sterilized soil from the excreta of various unspecified birds. The crops of wood pigeons shot in the west Midlands have been found to contain achenes of *R. repens* at certain periods of the year, notably in June–August, i.e. before ripe grain or stubble become available, with a sharp decline in September and none thereafter.

The general pattern accords with what Middleton & Chitty[50] found for the grey partridge (*Perdix perdix*): although they do not commit themselves to the species of *Ranunculus*, the photograph they give apparently includes *R. repens*.

Since birds frequently fall victim to predators, one method of dispersal additional to the shedding of excreta could come from the scattering of crop contents as their bodies are torn up.

Achenes of *R. repens* have also been found by Darlington in the winter food stores of certain small rodents, the field mouse (*Apodemus sylvaticus*) and short-tailed vole (*Microtus agrestis*): the former frequently, and the latter habitually, store food underground, a form of behaviour which might be expected to promote both dispersal and establishment.

Propagules distributed initially by one agency may be carried further by others. Ants take many seeds, notably those like *Veronica agrestis* and *Montia perfoliata* having elaiosomes attached which are particularly attractive to ants. Although propagules of *R. repens* appear to have no specialised structures of a corresponding kind, the ant *Lasius niger* has been noticed in autumn dragging the achenes over the ground.

2.4 Dispersal of asexual propagules

Seemingly those weeds having prostrate runners creeping over-ground are the forms able most rapidly to invade an area. For a creeping buttercup, 4 m² in a single season is acceptable as an approximation for the invasion of soil where the community remains open. We have recorded 7–10 m² on unshaded soil of boulder-clay-with-chalk from which potential competitors were removed by hand weeding, a value perhaps exceeded only by creeping cinquefoil (*Potentilla reptans*) for which an area of about 12 m² has been measured. Compared with this, ground elder (*Aegopodium podagraria*), spreading by underground rhizomes, dominates an area of only some 3 m² in a year under favourable conditions.

Rapid expansion of *R. repens* by its stolons may be contrasted with the very slow spread of michaelmas daisies and bulbous plants, where dormant buds at the base, protected by cover (leaf-litter, snow or soil), give rise year by year to new shoots and these, by decay of the parts connecting them with the original stock, to new plants which, because of their closeness, are able gradually to dominate adjacent ground. One advantage of spread of the latter kind over the method of exploitation by runners is that competitors are more likely to be eliminated by crowding out: stolons set up an open system which may enclose other species in its interstices without obliterating them.

A runner of *R. repens* has exceptionally long internodes; its tip exerts apical dominance in that, if removed, lateral buds behind it grow out. The initial horizontal extension of the runner is eventually superseded by vertical growths at the points of rooting, the vertical internodes being compressed, but the factors bringing about the change from the long internodes of the horizontal stage to the short ones of the vertical growths remain uncertain. Presumably control is under the influence of auxins, and presumably the stolons are diageotropic, whilst the ascending axillary growths, after rooting, are negatively geotropic and probably positively phototropic as well. Further work might include the application of auxins over a range of concentrations in lanolin: the work carried out already in testing the effect of light is summarised later.

Runners are homologous neither with tendrils nor roots, yet they show certain developmental affinities with both, correlated with the functions they perform. Tendrils commonly undergo increase in mechanical tissue on becoming secured to a support: there is similar development in runners after rooting. During elongation, runners make their way between surface obstacles somewhat after the manner of roots around impediments below the surface: at this time the pith is solid and the mechanical tissue of the runner is at first towards the centre, as it is in roots, an arrangement particularly suited to a sinuous pattern of extension without causing damage to the elongating structures. Following rooting of the adjacent nodes, the pith 'blows up' and finally breaks down to form a cavity; there is peripheral distribution of mechanical cells, and the runner is then better able to withstand lateral stresses, much as a vertical herbaceous stem can withstand buffeting by wind and moving animals. Fracture of the internodes decreases progressively in importance the

---

Fig. 2.4 STRUCTURE CORRELATED WITH FUNCTION. Sections through organs of *R. repens* which undergo elongation: a and b, with solid core of mechanical elements, are adapted for sinuous forms of extension between obstacles; c and d, with hollow centre and peripheral mechanical elements, are exposed to lateral stresses. For scale, see Fig. 2.1.

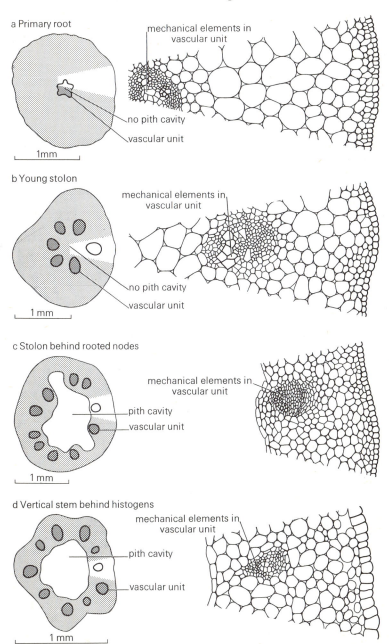

a Primary root

mechanical elements in
vascular unit

no pith cavity

vascular unit

1mm

b Young stolon

mechanical elements in
vascular unit

no pith cavity

vascular unit

1 mm

c Stolon behind rooted nodes

mechanical elements in
vascular unit

pith cavity

vascular unit

1 mm

d Vertical stem behind histogens

mechanical elements in
vascular unit

pith cavity

vascular unit

1 mm

nearer they lie to the parent stock, since separation by decay is the common development.

Fig. 2.4 illustrates some of these structural features.

Attachment of runners to the parent stock persists after growths at the nodes have become assimilatory and, if forcibly separated, self-supporting. Probably the most troublesome form of vegetative propagation is by those weeds wherein assimilatory shoots develop from fragmented roots. A number of plants with tap roots containing food reserves, such as horseradish and dandelion, exhibit this; and in carrot (*Daucus carota*) it has been found possible to grow entire plants in culture media from single cells of the root. On a larger scale, vegetative spread of creeping thistle (*Cirsium arvense*)

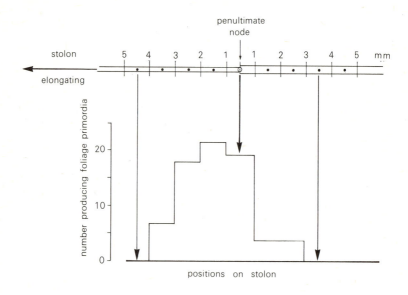

Fig. 2.5 TRAUMATIC DISPERSAL BY STOLON. Results of plating out on soil agar 200 discs, 1 mm thick, cut from within 1 cm of either side of a node on a *R. repens* stolon. All the 69 which produced assimilatory growths came from within 4·5 mm of the node. Development was greater from the younger than older inter-node: 66% came from this side, 28% from the node itself, and 6% from the older tissues. Evidently detached pieces of stolon, even where buds are not necessarily visible, may retain regenerative potential.

from detached fragments of various subterranean structures, has been considered in relation to disturbed arable land by Sagar & Rawson[65], and to unstable refuse tips by Darlington[18]. Experiments, both with artificial media and soil, indicate that creeping buttercup possesses the property of such traumatic dispersal to a limited extent. In cultures grown upon soil agar, assimilatory foliage has been produced out of runners both from buds cut from the stolons at nodes and from segments of internodes which appeared to be devoid of existing buds but which had developed within 4.5 mm of the nearest nodal axillary bud. Attempts to induce growths from adventitious roots have not succeeded.

The extent to which the localization of stored foodstuffs or of hormones can be accepted as responsible for such developments remains uncertain, and suggests material for further experimentation. Fig. 2.5 summarises the results of laboratory investigations on vegetative proliferation from detached segments of *R. repens* stolons.

Vegetational spread of the kind which commonly occurs in creeping buttercup results in colonization of nearby areas in contrast with a scattered form of dispersal, and gives rise to clones wherein the descendants have the hereditary composition of the parent stock without the advantage of genetical variation. Were asexual reproduction to be the sole method, presumably a species would be liable to eventual extermination.

Material from creeping buttercup was used for comparing effects of clonal derivation with those of sexual fusion. One example is presented here, where development of a diagnostic feature of *R. repens* – the long midrib stalk supporting the terminal leaflet of the foliage – is compared in two widely separated clones and in scattered plants originating from a general gene pool. One hundred fully expanded leaves, collected at random, were examined from each of three situations: a clone on damp, clayey soil in a shaded locality in Worcestershire; a second clone on lighter soil in well-illuminated surroundings at the Malvern College Nature Reserve in Herefordshire; and material taken from a variety of conditions in the west Midlands.

Fig. 2.6 shows the method of standardisation: a hypothetical assimilatory area is determined as the product of two maxima, the length (l) and the width (w) of the leaf: the length of the stalk (s) is taken as the measurement from the proximal margin of the terminal leaflet to the union of the posterior leaflets with the midrib.

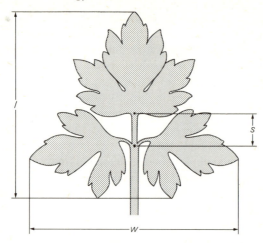

Fig. 2.6 STANDARDIZATION OF LEAF MEASUREMENTS. Method used for comparing *R. repens* foliage taken from plants differing in heredity and environment. The letters refer to the formula in the text.

Comparisons are derived from the formula

$$\frac{s}{l \times w} = mm^{-1}$$

Notwithstanding considerable variation in actual leaf-size within a clone the method gives significant comparisons between materials which, at a glance, seem dissimilar. Thus, two of the leaves near the peak value for one of the clones measured were respectively

l 25 mm, w 25 mm, area 625 mm², s 2 mm, $\dfrac{s}{l \times w} = 0.0032$ mm⁻¹

l 48 mm, w 50 mm, area 2400 mm², s 8 mm, $\dfrac{s}{l \times w} = 0.0033$ mm⁻¹

Comparisons are shown graphically in Fig. 2.7. All give typical distribution curves, the peak differing in the three, and the clones having a much smaller variation than material from the open gene pool.

Differences between growths arising from clones and those of more general origin could be the results of factors other than those inherited: the limited range, of environmental conditions, for

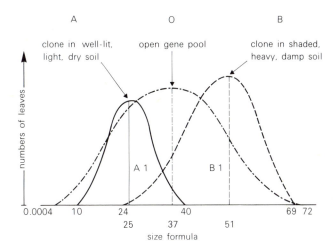

Fig. 2.7 ENVIRONMENTAL AND GENETICAL EFFECTS. Distribution curves of the sizes of *R. repens* leaves taken from two clones and a random sample. Note the narrow range in each clone. The data used are shown in Fig. 2.6. West Midlands.

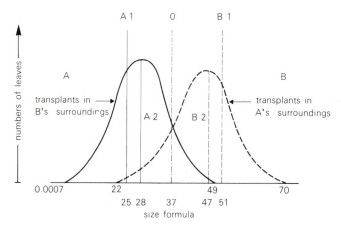

Fig. 2.8 ENVIRONMENTAL AND GENETICAL EFFECTS. Results from measurements of *R. repens* foliage of a transplant experiment. Compare with Fig. 2.7. Although there is an extension of the two clones and a shift of their peaks towards that of the open sample, they retain their identity. West Midlands.

41

example, to which clonal offspring tend to be subjected, certainly in *R. repens*. Assessment of the relative efficacy of such factors clearly involves elimination of the differentials. In the two clones under discussion, transplant experiments were carried out effecting an interchange of detached nodes, those from each clone being set up in the surroundings of the other before roots or leaves had developed. Measurement of foliage eventually arising from the transplant are compared in Fig. 2.8 with the original values: although there is a small shift towards the open samples both of the peaks and extremes, the general character of the original clones remains and suggests the importance of the genetical composition. Whether the shifts indicate the consequences of actual disturbance or imply edaphic and climatic influences are problems for further experimentation lending itself to comparatively straightforward projects within the capabilities of resourceful sixth-form students.

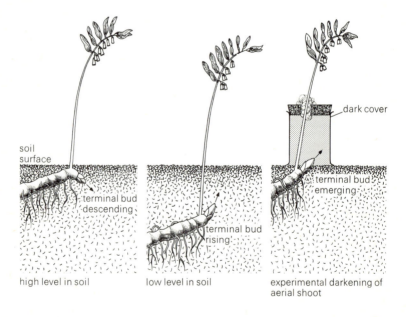

high level in soil   low level in soil   experimental darkening of aerial shoot

Fig. 2.9 ILLUMINATION AND STEM-GROWTH. Effect of darkening the aerial stem in Solomon's seal (*Polygonatum multiflorum*) on the subsequent direction of growth of the underground rhizome. Adapted from Raunkiaer[63].

2.5 Effects of light

The diversity of situations colonized by creeping buttercup points to a wide tolerance of environmental variables, including those of lighting. Commonly it is part of the community of relatively unshaded grassland.

Effects of illumination upon the growth of such horizontally orientated stem structures as rhizomes has already been investigated to some extent, notably by Raunkiaer[63], who found that the adjustment in depth, by further growth of the subterranean rhizome of Solomon's seal, was influenced by the extent to which the vertical overground flowering stem was experimentally darkened (Fig. 2.9). Phillips[59] in his chapter on apical dominance discusses this concisely. Raunkiaer was concerned more with the maintenance of the rhizome's depth in the ground than with quantitative increase in length, and there is scope for additional work involving other species on the reactions of corresponding structures to changing physico-chemical conditions as well as to climatic variables.

Observations frequently suggest a close connection between light and the horizontal spread of specialized stems adapted for the vegetative conquest of nearby spaces. Thus, the rooting branches of bramble and other rosaceous plants elongate with impressive rapidity in shade; much the same happens when the herbs on the floors of woods and hedgerows spread by overground runners; and the stolons of creeping buttercup show a similar trend when the parental growths occupy situations where the illumination is low. That light influences the elongation of internodes is clear from the reaction of a buried dandelion plant, which grows rapidly to the surface and there forms a new rosette. Shoots of plants like marram-grass develop similarly when buried under the accumulations of enlarging sand-dunes: both marram in dunes (Fig. 9.1), and coltsfoot in refuse heaps and the mounting debris from landslides (Fig. 9.3), produce new photosynthetic growths which keep pace with successive rises in the surface level. Creeping buttercup is among the most successful of the colonizers of such disturbed places, and the burial of the nodes of its spreading stolons seems typically to bring about an increase in the clusters of assimilatory foliage appearing at the new surface. Whether the spread is entirely horizontal, or whether its direction also has a vertical component, one effect is clearly that of transferring new growths to conditions where some of the factors may be more favourable for establishment than in the immediate surroundings of the original stock.

2.6 Relationship between light and growth

Experiments carried out on the effects of burial and of shading without burial showed two aspects of the apparent relationship between light and growth in *R. repens*.

Growths stimulated by burial were observed by planting detached leafy nodes in deep boxes and, after they had become established, covering them with finely divided soil. Effects included marked separation of the nodes, normally very close, in the upright stems which developed, and the production of axillary runners which first elongated at an angle of about 30° with the vertical and then, on reaching light, arched downwards until contact was made with the surface and thereafter extended horizontally (Fig. 2.10). Shading was also brought about by inverting wooden tubs over rooted nodes in the field and periodically irrigating the growths capped by them. Results (Fig. 2.11) can be summarised as follows:

1 production of internodes which were widely separated on the sections of runners in the dark and succeeded by shorter internodes after the runners had passed into the light;
2 an increase in the actual number of runners spreading radially from growths which were capped;
3 no increase in runners from proximal nodes where only the distal nodes were shaded;
4 no effect on developments from distal nodes where proximal nodes alone were darkened.

Doubtless auxins are operative here, but their effect seems to be more localized than in vertical stems, and to correlate with the early independence of growths which, on separation, can give rise to new individuals.

2.7 Effects of water

That the balance between asexual development and reproduction by seed may be affected by water in terms of climatic variants is suggested in observations made by Salisbury[66] on yellow archangel (*Galeobdolon luteum*). He found in two successive summers, the first

In light

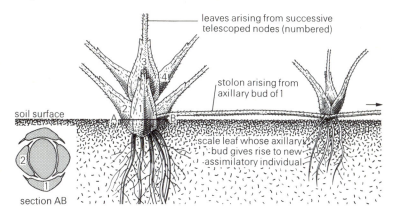

leaves arising from successive
telescoped nodes (numbered)

stolon arising from
axillary bud of 1

soil surface

scale leaf whose axillary
bud gives rise to new
assimilatory individual

section AB

Parent buried

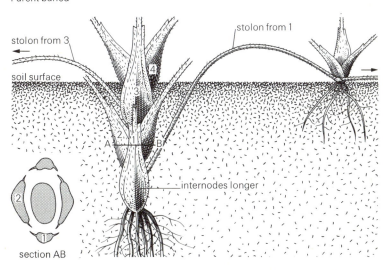

stolon from 3

stolon from 1

soil surface

internodes longer

section AB

Fig. 2.10 ILLUMINATION AND STEM-GROWTH. Internodal extensions
from two assimilatory units of *R. repens*, one buried experimentally under shallow
capping of soil.

One Approach to Ecology

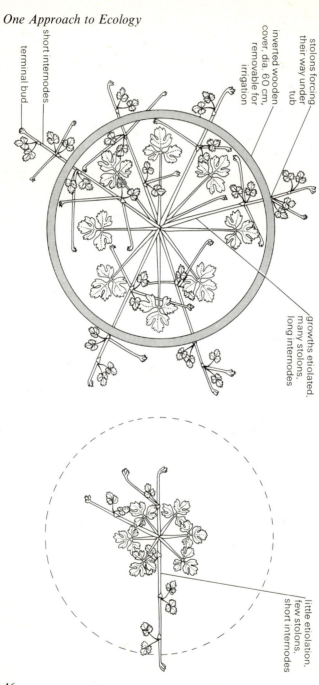

stolons forcing their way under tub

inverted wooden cover, dia. 60 cm, removable for irrigation

short internodes

terminal bud

Darkened

growths etiolated, many stolons, long internodes

Illuminated

little etiolation, few stolons, short internodes

46

dry and sunny and the second dull and wet, that the same plants growing in partial shade flowered freely in the dry summer, but produced fewer runners (3–4 per plant) than in the following season (4–5 per plant) when, in addition, the number of fruits set was roughly halved. He also found that the runners formed in the wet summer were longer and bore a higher number of rooting nodes. But the terms 'wet' and 'dry' applied to seasonal conditions involve differentials other than those of moisture, and the factors which actually brought about the results remain uncertain. No information on this topic is available for creeping buttercup, so that the need for further work is self-evident.

That the amount of water in the substrate is important, and possibly decisive, in allowing creeping buttercup to gain a foothold at a particular site, appears from the comparisons made by Harper & Sagar[34] with the other grassland buttercups and from comparisons made by Darlington[18] with weeds of other genera at a refuse dump.

Records over a period of some two years at a large compost heap in a Worcestershire garden took account of the development of several weed species which had arrived either with the waste itself or with the soil used to bury it. After being treated with activator the pile was finally capped with untreated, weed-infested soil of a loamy consistency in the autumn of the first year, at which time it was conical, 1.75 m high at the apex and 3.5 m across the base: thereafter it was left undisturbed for 21 months during which time it underwent subsidence without losing its overall shape. Nearly two dozen species of vascular plants increased in biomass after arrival, and of these measurements were made in three: creeping buttercup, the only true perennial; scarlet pimpernel, *Anagallis arvensis*, and annual meadowgrass, *Poa annua*. The latter two, both typically annuals, were capable of persisting as biennials, when *Poa* was seen to produce small rooting branches whereas *Anagallis* appeared not to spread vegetatively despite its procumbent habit. Post-seedling plant structures became scattered at random over the surface of the capping soil: although many of these were fragmented they included segments which were either rooted already or had rooting potential. The highest proportion of rooted clumps surviving to undergo

Fig. 2.11 ILLUMINATION AND STEM-GROWTH. Stolon-development during five weeks from similar stocks of *R. repens*, one darkened experimentally.

further development was in the buttercup, where it was about one half; in *Poa* survivors were about one quarter and in *Anagallis* one seventh. Emergence of the first seedlings of all three species came within a fortnight of capping; a result at variance with the Harper-Sagar findings under different field conditions, where all the *R. repens* seedlings appeared in the late spring. Conditions at the heap probably broke the dormancy of buried seeds.

Both seedlings and older plants of *R. repens* became established mainly on the lower levels, in that part of the substrate which remained so wet as periodically to be waterlogged, colonization being densest near the angle between the slope and the horizontal surface of the clayey soil on which the pile had been built. On the higher and drier parts of the mound most plants in any stages of post-germination died within a month. During the growing season of the second year, stolons emanating both from first-year and older plants extended slowly up the sides and more rapidly over the level ground nearby, rooting at intervals and producing assimilatory structures. By the spring of the following year *R. repens* had become dense around practically the entire lower cone to within 35 cm of the crest, even in places where its pioneers earlier had failed to survive. At the higher levels it competed with *Anagallis* and largely eliminated it.

Broadly, the sequence followed that outlined by Lieth[43], who found that *R. repens* and also hop trefoil, *Trifolium campestre*, originated from small areas under near optimal conditions for establishment and that vegetative expansion then took place from these into areas which were not optimal. Lieth was considering only those plants developing from seeds which arrived by chance and not, as here, those developing from partly grown specimens deliberately concentrated in our locality by human agency; and he was concerned neither with *Poa* nor *Anagallis*.

2.8 Relationship between substrate water and buttercups

Harper and Sagar[34] recorded at an Oxfordshire locality the distribution of the three perennial buttercups in permanent grassland on ridges and furrows where the water content of the substrate tended to be maximal in the troughs and minimal on the tops of the ridges. As the particular site had been grazed for at least 800 years, the communities therein were supposedly stable. They found the

highest number of *R. repens* plants occurred in the drainage hollows
and the lowest number along the elevated crests where the water was
minimal; conversely *R. bulbosus* reached its peak in bands along the
crests was absent altogether from the bottoms of the channels; and
plants of *R. acris* had an intermediate status, growing in all situations but mostly on the sides of the slopes, with minima on the ridges
and in the furrows alike.

Using pasture of unknown history in a region with a drier
climate near the coastal town of Walton-on-the-Naze, Essex, the
writer prepared two belt transects, one running north-south and the
other east-west, across a hollow in an otherwise level field. The
depression, a drained pond, still held standing water for a time after
heavy rain. Results are charted in Fig. 2.12 and include three features
which are particularly noteworthy: there is confirmation in general
terms of the Harper-Sagar findings, although *R. bulbosus* proved to
be the only buttercup over extensive stretches of the horizontal field
where the ground was dry, with *R. acris* largely restricted to the slope
and *R. repens* mainly to the floor of the old pond; in these conditions, *R. repens* appeared to have made little headway in competition with established colonizers by vegetative expansion from
sites of primary occupation – in contrast with its development at the
refuse heap noted earlier where, presumably, competition pressures
were less; and the influence of aspect was apparent in restricting
growths of *R. repens* on the south-facing slope, where the drying
effect of the sun was probably operative. Under conditions of
variable drainage, therefore, the three species can be regarded as
fitting neatly into distinct niches in the environment, *R. repens* giving
the impression of being the most tolerant, and *R. bulbosus* the least
tolerant, of water-logged conditions.

But the relationship between water and the perennial buttercups
was further studied in the Harper-Sagar investigation by transplantation and germination experiments.

Rooted photosynthetic growths of all three species were transferred to pots in which the water tables were controlled, the first
series being allowed to drain from the bottom of the pots, the
second having a water table controlled at 4 in from the soil surface,
and the third with a water table maintained at the surface itself.
One effect was the extension of the flowering period in the transplants beyond that observed in the field, although *R. repens* flowered
less freely than the other two in the transplants, a feature possibly

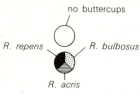

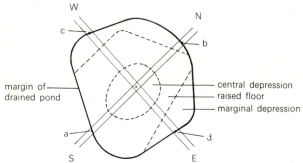

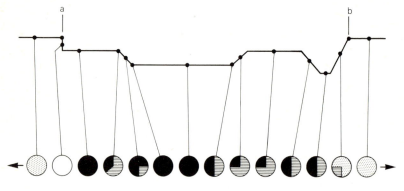

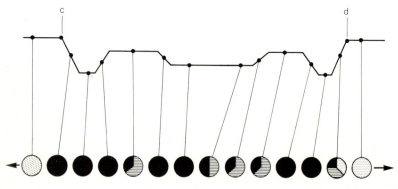

linked with the extensive production of stolons in the potted plants. The experiment indicated that all three could tolerate waterlogged conditions and survive therein, the dampness of the substrate mainly promoting the production of flowers: this leads to the supposition that water itself does not influence the suitability of a situation for *mature* plants of any of the three buttercups as much as might be thought to be the case from general field observations. It accords with the observed invasion of dry situations from wet initials on the part of creeping buttercup established at a refuse dump and noted earlier.

However, the influence of the water table on the germination and establishment of *seedlings* appeared to be decisive. Sowings of seeds of the three buttercups were made under conditions of varied water tables similar to those used for the transplantation experiments, and both Harper-Sagar and the authors sowed mixed populations of the three species of seed in the water-table variants. Emergence results of our experiment are shown in Fig. 2.13: clearly the relationship between seedling establishment and the water table is precisely what would be expected from the observed distribution of plants in the field. Furthermore, in a dry year, Harper and Sagar noted marked differences in the times of germination in the field: *R. bulbosus* made rapid germination in the autumn with negligible germination in the following spring; all the *R. repens* seedlings emerged in late spring; and *R. acris*, the species occupying the intermediate of the three niches, showed two distinct phases, an autumn germinating and a spring germinating phase. Our observations have revealed a similar trend, although the groupings were less clearly defined.

The initial advantage gained by a species through establishment of seedlings gives it an advantage over associated species which is maintained in later development. This is particularly true of *R. repens* which, on first becoming established, is then vigorous at all

---

Fig. 2.12 GRASSLAND BUTTERCUPS AND WATER. Ratios between the cover of three species along transects crossing a drained pond. Length a–b = c–d = 11 m. Measurements of cover value were made with a point quadrat frame but similar results appeared when line transects were set up and records taken of the buttercups touching these. *R. repens* appears to have been influenced both by hollows in the ground and the sun's direction. Walton-on-the-Naze, Essex.

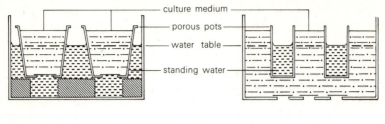

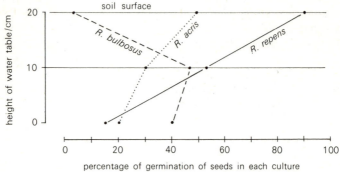

Fig. 2.13 GRASSLAND BUTTERCUPS AND WATER. Emergence of seedlings under water-table differentials: *upper*, simple methods of controlling height of water table in laboratory cultures; *lower*, germination successes from three groups of cultures.

water-tables and starts invading adjacent ground vegetatively within a couple of months of sowing.

In view of this explanation, which seems to stand up to experimental tests, for the observed distribution of *R. repens* relative to ground slope, it is fitting to consider some of the other plants in short grass which also appear to be distributed unevenly in accordance with variations in height of the ground surface. Among these is white clover, *Trifolium repens*, another creeping perennial rooting at the nodes and one which seems to be particularly sensitive to such variations. A suggestion which has commonly been advanced here is that height may so affect the illumination available to the foliage of a light-demanding but low-growing colonizer that this is probably the causative factor: but it is evident from the foregoing that hypotheses of this kind require experimental support before acceptance.

Creeping buttercup's close association in its early phases with the substrate moisture is seemingly accompanied by a somewhat unexpected feature: under certain conditions of saturation in laboratory cultures its seedlings at the three-leaf stage appear to be unique among those of the grassland buttercups in undergoing guttation, droplets of water forming, often copiously, along the leaf margins despite the fact that hydathodes have not been noticed here. We have found this phenomenon so erratic and unpredictable– somewhat after the manner of root pressure–as to suggest that the circumstances in which it occurs require further study.

2.9 Some conclusions

The wide distribution of *R. repens*, and the range of situations and soils in which it can establish itself, point to the adaptability of a species which is particularly effective as a weed. Some of the factors limiting its establishment are inferred in the three aspects of colonization outlined below.

1 Despite the relatively sparse production of flowers–a feature sometimes put forward by ecologists as one correlated with a well-developed method of asexual propagation–the setting of seeds and their dispersal are demonstrably important in effecting initial establishment at a critical phase in the life history, when the substrate requirements necessitate conditions of near-saturation and from which vegetative expansion can take place into less restrictive situations. The sexual bodies are carried by such agencies as wind and moving solids and the need for these to operate is self-evident.

2 Extension from a site of primary occupation may bring about competitive relationships in which, despite its stoloniferous form and production of phytotoxicant, creeping buttercup is at a disadvantage. This is sometimes apparent in the drier parts of permanent grassland where peripheral growth from a wet, low-lying initial is inhibited by vegetation already established. Where the soil is freshly disturbed, invading *R. repens* seems to have the advantage over some of the annual or biennial pioneers.

3 Being the least poisonous of the grassland buttercups, *R. repens* would seem to be placed at a disadvantage where there are grazing animals, although fragmentation can bring about re-establishment.

2.10 Key to British plants with buttercup-like flowers

This key distinguishes between species which bear yellow flowers looking ± like open saucers and in which some at least of the floral parts are spirally arranged. It includes the genera *Caltha* (marsh marigold) and *Trollius* (globe flower) as well as the rarities *Ranunculus flabellatus* (Jersey buttercup), *R. ophioglossifolius* (adders-tongue spearwort) and *R. reptans* (creeping spearwort) whose distribution is restricted to exceedingly narrow geographical limits. It omits very common and widely distributed species of *Ranunculus* with white flowers, e.g. *R. aquatilis* agg. (water crowfoot).

	Perianth of 5–15 large, yellow, petaloid segments without green sepals	1
	Perianth of green sepals and yellow petals	2
1	Leaves simple with toothed margins and long stalks *Caltha palustris* (marsh marigold) Widespread in marshy places	
	Leaves deeply cut and arranged spirally *Trollius europaeus* (globe flower) Gullies and damp pastures in hill districts in north	
2	Perianth of 3 sepals and 7 + petals *Ranunculus ficaria* (lesser celandine) Widespread in dampish, shady places	
	Perianth of 5 sepals and 5 petals	3
3	Leaves cut or lobed	4
	Leaves simple	12
4	Fruits smooth or with small warts	5
	Fruits spiny or with hooked hairs	11
5	Sepals folded sharply back in open flower	6
	Sepals ± straight in open flower	7
6	Stem with swelling (corm) at base; fruits smooth *Ranunculus bulbosus* (bulbous buttercup) Widespread in drier grasslands	

Stem without corm; fruits with a few warts
Ranunculus sardous (hairy buttercup)
Local in grassy places, especially near coast

7   Leaves ± glabrous   8
     Leaves hairy   9

8   Flowers small (0.5–1.0 cm dia); fruits in oblong head
     Ranunculus sceleratus (celery-leaved crowfoot)
     Widespread in muddy situations

     Flowers large (1.5–2.5 cm dia) but often imperfect;
     fruits in spherical head
     Ranunculus auricomus (goldilocks)
     Widespread but local in woods or on rocks

9   Stem leaves 1–2, small; tubers at base of stock; flowers
     large (2.5–3.0 cm dia) with very glossy petals
     Ranunculus flabellatus (Jersey buttercup)
     Very rare, on hot dry banks in Jersey

     Stem leaves 2 +, lower ones large; no tubers   10

10 Long, prolific rooting runners; flower stalk furrowed
     Ranunculus repens (creeping buttercup)
     Widespread in moister situations

     No runners; flower stalk not furrowed
     Ranunculus acris (meadow buttercup)
     Widespread in meadows and pastures

11 Sprawling plant, diffusely branched; flowers small
     (6 mm dia); fruits very rough
     Ranunculus parviflorus (small-flowered buttercup)
     Widespread but local on dry, calcareous soil

     Erect plant with erect branches; flowers large (1.2 cm dia):
     fruiting heads like burs
     Ranunculus arvensis (corn buttercup)
     Widespread but scarce cornfield weed

12 Tall plant (60 + cm); flowers large (2.0 + cm dia)
     Ranunculus lingua (greater spearwort)
     Local in marshes, ditches and fens
     Short plant (under 60 cm); flowers small

(under 2.0 cm dia)   13

13 Erect plant with broad, spoon-shaped basal leaves;
    small flowers (6–9 mm dia); warty fruits
    Ranunculus ophioglossifolius (adderstongue spearwort)
    Very rare and confined to two marshes in Gloucestershire
    Decumbent, creeping or ascending plant; basal leaves
    not spoon-shaped; flowers up to 18 mm dia; fruits
    without warts   14

14 Creeping stems rooting at every node and arching in
    internodes; flowers 5–10 mm dia; fruits with long
    terminal hook
    Ranunculus reptans (creeping spearwort)
    Very rare and confined to 1–2 gravelly loch-sides

    Creeping stems not rooting at every node; flowers solitary
    or few; fruits with very short beak
    Ranunculus flammula (lesser spearwort)
    Widespread in ditches, marshes and by sheets of water

3 The habitat

3.1 What is a habitat?

The previous chapters have examined some of the ways in which two selected organisms are adapted for their respective environments.

Simply expressed, the habitat of any organism is the place where it lives. In fact, the habitat is the smallest part of the environment wherein it can shelter, feed and reproduce. Shelter is from climatic, edaphic and biotic conditions detrimental to survival. Feeding includes the acquisition of water, either directly or from metabolic processes. Reproductive success becomes significant only when it results in the replacement of breeding individuals lost through death: reproduction, the ultimate test of establishment in a habitat, is an expression of the success in obtaining shelter and food.

Stated in these terms, the concept of a habitat as something with a defined form and delimited boundaries is untenable. Seashores, sand dunes, heathland, moorland, woodland, etc. support populations of animals and plants living in diverse habitats which together form an ecosystem. Space, as such, must be considered in relation to a particular organism. For example, how is the habitat of a migratory bird to be conceived; and does the habitat of the rotifer *Callidina symbiotica*, which lives practically confined to the tiny reservoirs at the bases of leaves of one species of liverwort, *Frullania dilatata* (Fig. 3.1), qualify as a 'microhabitat'? Indeed, when is the term microhabitat valid as denoting anything distinctive?

3.2 The more restricted habitats

A microhabitat is not necessarily one which is markedly restricted in space. Conversely, for some organisms, comparatively large habitats are restricted if the pressures they exert on their occupants are heavy when these living things are in process of adaptation to specialized conditions.

The olm, *Proteus anguinus*, Fig. 3.2 is an example of adaptation to somewhat extreme conditions. The animal is found only in the total

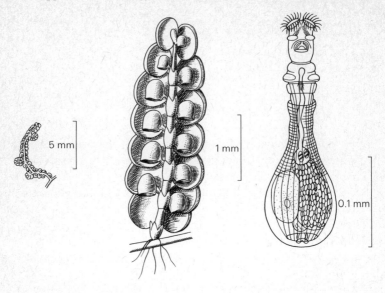

Fig. 3.1 RESTRICTED HABITATS. Rotifer, *Callidina symbiotica*, found in the reservoirs on the fronds of the liverwort, *Frullania dilatata*.

darkness of cold underground streams, notably in Dalmatia. A colourless amphibian, virtually blind and with almost completely atrophied legs, it possesses external gills and lies motionless in the water from which it derives oxygen and food located by the lateral line system. In other words the olm, by severe reduction of limbs, loss of pigmentation and eyesight, appears to be completely adapted to a troglodytic existence.

Modification of habit as well as of structure may be required. The marine iguanas, *Amblyrhincus cristatus*, Fig. 3.3, which are unique to the Galapagos islands, live on the coastline, dividing their lives between basking on the lava rocks and foraging in the sea. These lizards, descendants of terrestrial forms in South America, have evolved a method of swimming by holding their forelegs tightly to their sides, trailing their back legs behind and using their tails as rudders to propel them through the water. Their blunt snouts enable them to browse off seaweed growing plentifully down to about 10 metres below the surface and whilst below, they can drink sea water eliminating the excess salt by means of special salt glands. On shore they maintain a body temperature of 35°C or more, yet since these

external gills

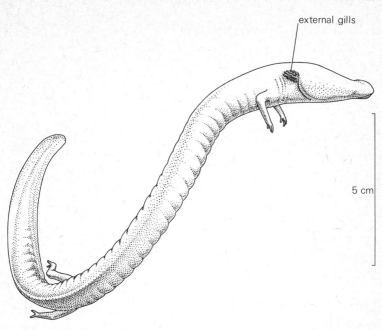

Fig. 3.2 RESTRICTED HABITATS. Olm, *Proteus anguinus*, approximately 30 cm long. An unpigmented, neotenous amphibian, showing extreme degeneration of limbs and eyes in response to a troglodytic existence in almost total darkness.

marine iguanas, like all reptiles, are poikilothermic, their body temperature, when in the sea, quickly falls to that of the surrounding water which may be only 26°C.

The scarcity of accessible land vegetation on the islands may have been one of the chief reasons for the adoption of a partially marine existence by these reptiles, compelling them, whilst on their submarine search for seaweed, to operate at a lower temperature than the preferred temperature. Their physiological adaptations do not permit them to remain beneath the surface for very long, for as soon as their body temperature falls below a critical point they must return to bask on the rocks and 'recharge their batteries' so to speak. While in the sea they possess little defence against their natural predators, the sharks, and on land are confined to the narrow strip of spray-washed lava rocks bordering the sea.

Restricted habitats can impose extreme morphological and physiological modifications. Examples are to be found among many of the internal parasites. One of the most bizarre is the marine copepod, *Xenocoeloma brumpti*, described in Section 4.2.

Fig. 3.3 RESTRICTED HABITATS. Marine iguana, *Amblyrhincus cristatus*, approximately 120 cm, found on the Galapagos Islands.

3.3 The less restricted habitats

Two examples are given.

The Swift
In connection with those species whose habitats are inconceivable as circumscribed entities, the swift (*Apus apus*) merits consideration as a common, highly specialized and far-ranging bird. Exclusively aerial, all its activities take place on the wing or at the nesting site. It does not voluntarily settle on the ground, its short legs being adapted only for clinging to rocks, but it can rise from a level surface. On occasion it appears over terrain of almost any sort, from lowlands to mountain tops. During the breeding season its local abundance appears to be determined by the availability of food and suitable nesting sites.

Originally a breeder in such restricted situations as hollows within sea cliffs and upland crags, the swift is now so closely associated with the multifarious recesses within large buildings in towns as to become predominantly urban. Some cities in southern Europe have very large populations. In assessing its affinity for towns a subordinate factor is its tendency when on passage to remain for a time circling near buildings with potential nesting places, whether or not these are actually occupied at the time, so that the number frequenting an urban situation may temporarily exceed the population remaining to breed.

Dipterous insects make up the bulk of the young swift's food. They are taken by the parent birds while on the wing, the birds being capable of roaming far when feeding. The insects so collected are rolled into a sphere in the throat before being expelled as a pellet into the mouth of a nestling. Cool, rainy weather brings about such a decline in airborne material that it is usual, in such conditions, to find great assemblies of swifts circling low over sewage farms and some kinds of refuse both inside and beyond urban areas. Presumably these are places from which two-winged flies, such as chironomids which emerge throughout the whole of the 14 summer weeks when swifts are around, are taken in abundance. Though attractive at all times such places are especially so when there is a deficiency of food over less specialized situations. Feeding movements, particularly when young are being tended, can result in concentration of swifts from a large area at a favoured locality.

Adaptation to a climate whose weather periodically inhibits flight of the air-borne insects constituting the exclusive food of the young birds, includes their ability to survive intermittent feeding. Intervals when they are deprived of food result in a fall both of body temperature and weight, the nestlings becoming almost poikilothermic until renewed feeding induces a temperature rise and the resumption of growth. Such birds grow in a series of steps under conditions which might well prove fatal in the absence of such adaptations. Varying estimates of the time spent by the young in the nest range from 35 to 56 days and reflects the irregularity of feeding and growth.

Swifts can be said to have become adapted to the new conditions of urban development by modifying that part of the habitat where breeding takes place. Had this not happened doubtless they would have survived by continuing to use nesting sites of the original sort unaffected by urbanization, the likely consequence of which would have been a widespread and diffuse population, low in numbers.

61

The Red Deer (See Darling[16])
More striking still is the case of the British subspecies of red deer,
Cervus elaphus scoticus, formerly an animal of extensive, thick
woodland, which in our part of the world is now found principally
in habitats of a very different kind. These are the bare open moor-
lands of the Scottish Highlands and islands called 'forests', with
remnants in Westmorland, Devon, Somerset, the New Forest and
County Kerry in Ireland. Had this beast been restricted to its
earlier type of habitat, it might well have become extinct due to the
clearance of the Caledonian Forests following the 1745 rebellion
when moors took the place of the afforested regions. As things are,
the species is able to survive in the new conditions provided that,
at certain periods of the year, it has access to cover.

Red deer are gregarious but the sexes live apart except during the
rutting season, groups living in definite territories from which they
do not normally stray. These territories are comparatively small but
those of different groups commonly overlap. Mild-weather and
hard-weather territories differ, the former being higher on the
moors and the latter at the lower, more sheltered levels. In winter
the stags are solitary while the hinds and young of one and two
years, are in herds of 10 to 40 in their usual territories. Each spring
the stags drop their antlers. Replacements start to grow immediately
and most stags are driven to the high ground by flies since the 'velvet'
covering the new horns is very sensitive. Here they collect in herds
of 3 to 20 or more, although some continue to hide in the woods
until rut. Stags keep to their herds up to mid-September when they
break up to seek out the hinds. A vigorous stag normally gains
possession of a herd of hinds, to which he may add any others he
can collect, eventually acquiring a 'harem' of 100 or more and
defends them against rival males.

Towards late May the hinds separate from their herds to give
birth to calves, making use of such cover as is available and joining
up again into herds as soon as the calves can follow.

Herds feed in the evening and before dawn, returning to their
territories in the early morning where they spend most of the day
doing little except chewing the cud. Red deer need for resting, a sheltered, sunny place in their territory where two winds meet and they
have a clear view, their sense of smell being so keen as to enable
them to detect human scent over a distance of a mile or so in favourable conditions. They are so sensitive to weather-change as to leave

the moorland summits before a storm and to face out the snow in the open, windswept terrain instead of being caught in drifts.

Change of environment has presented few problems in feeding to so general a herbivore, which eats a variety of plant material ranging from grasses, mosses and heathers on the open moor to seaweed on the shore and leafy shoots in woods.

Despite its ability to adapt to new conditions created by man, its survival into the present century owes much in England to hunting and in Scotland to stalking. Much the same is true of foxes and fox hunting. Although the subject is so beset with emotional overtones as sometimes to make objective discussion difficult, this kind of human activity is perhaps unexpectedly beneficial.

4 Relationships and interdependence

4.1 The concept of community

Like 'success', the word 'community' holds different meanings for different people, a diversity which makes it less useful for defining an ecological entity. To us, a community is made up of organisms sharing the same environment and the term implies relationships. Any ecological system which is simple tends to be unstable, and an animal community can become so complex as to make for difficulty in assessing the correlation between its members and their environmental conditions. Since the difficulty arises partly from an animal's mobility, how can the communities of far-ranging birds be delineated? The swift has already been considered in reference to some of the spatial complexities involved (3.3).

4.2 Close relationships

Consideration of habitats and communities includes consideration of those cases where one living organism provides habitats for others and, so long as it remains alive, sustains them. These close relationships, forming a graded series in respect of intimacy, are commonly designated by such terms as parasitism, symbiosis and commensalism.

Parasitism
Parasitism (see Croll[15]) is a development from the food-chain relationship of free-living organisms considered in the next chapter and generally requires survival of the host for the attacker's survival. While it is true that a parasite usually harms its host in some way, only in extreme cases does it encompass death. Should this happen, the *coup-de-grâce* is delivered when the host's death does not operate to the attacker's disadvantage: for instance, those diptera and hymenoptera which, as larvae, feed endoparasitically within

aphids and caterpillars and kill their hosts, have completed feeding and are ready for pupation when death occurs. Were the relationship otherwise, it would be a case of killing the goose that lays the golden eggs since the attackers are obligate parasites unable to feed on dead remains.

A common parasite of this kind is the brachonid *Apanteles glomeratus* which attacks the younger larvae of both the large and small white butterflies (*Pieris brassicae* and *P. rapae*). A female *Apanteles* pierces the epidermis of a first or second instar caterpillar and may deposit over a hundred eggs: the larvae from these feed on the fat-body but avoid the host's nervous system. At the stage when it would normally pupate the infested caterpillar dies: the *Apanteles* larvae emerge from the still-living remains and spin yellow cocoons nearby. Among the interesting features of the relationship is that the *Apanteles* larvae are not themselves safe from attack when inside their host and, in fact, are often parasitised by a small ichneumon, *Hemiteles fulvipes*, the females of which reach them with their ovipositors. Emergence of *Hemiteles* from the cocoons instead of *Apanteles* is an observable result. This is not the end of the sequence, for the larvae of *Hemiteles* in turn can be parasitized by a small chalcid wasp of the genus *Pteromalus*.

Thus there are examples in this chain of events of primary parasitism of the cabbage white caterpillar by *Apanteles*, secondary parasitism by *Hemiteles* which does not parasitize directly but through the primary parasite, and tertiary parasitism by *Pteromalus*. Perhaps this type of association, where the host is normally destroyed, should not be regarded as true parasitism but as a successful form of internal predation, more accurately described as parasitoidal. Organisms such as these, which are so adapted for the parasitic mode of life as to secure for themselves both shelter and a continuous supply of food just so long as they need them, can be regarded as highly specialized predators which budget well. Facultative parasites, like those fungi which can persist as saprophytes on dead organic matter when suitable living hosts are not colonized, budget particularly well.

Although there are many endoparasites, such as tapeworms, which cause deleterious effects including starvation, debility and nervous disorders, a host and its parasite sustain as a rule so nicely balanced a relationship that it is exceptional for the reproductive capacity of the host–the ultimate test of biological success–to be stopped altogether. Thus, oak trees are commonly seen to be thickly infested with galls which are their response to attacks by certain kinds of parasitic insects, in the main bugs, midges and cynipid wasps. Many of the galls contain entire habitats wherein live communities of remarkable complexity, made up of parasites, hyperparasites, predators, inquilines and successori. Yet, despite the deflection of materials from the plant's metabolic resources to localized sites where these dependents make use of them in sundry ways, there is no decisive impairment of the main host's reproductive potential and it continues to set seed in quantity (Darlington[17]).

The parasitic association of the marine copepod *Xenocoeloma brumpti* and its host, the terebellid *Polycirrus arenivorus*, affords an example of extreme specialization on the part of the parasite to the point where the marked reduction of the body makes it almost unrecognisable as a member of the order Copepoda. Although morphologically an external parasite of the terebellid, *Xenocoeloma* has become, in reality, an internal parasite. After the parasite has become attached to the body wall of the host, the tissues of the latter, like those of plants infested by gall-causing hymenopterans, grow to enclose *Xenocoeloma* completely, except for a single terminal aperture to the exterior through which the ovaries discharge their products. The epidermis of the parasite disappears and the animal becomes covered by the host's epidermis which, in turn, becomes lined with the parasite's muscles. These various layers enclose a space, the axial cavity, lined with the host's endothelium. The parasite possesses no digestive tract, nervous system or appendages, and is hermaphrodite. Nourished by its host's coelomic fluids, *Xenocoeloma* has attained the ultimate in parasitism by incorporating within its own body part of that of its host.

Although they may be vectors of bacterial or virus infections, ectoparasites in general, like fleas and ticks, are less injurious to their hosts than are endoparasites. Many organisms commonly regarded as endoparasites, but which cause no apparent injury, are better regarded as saprozoites. The ciliates *Balantidium* and *Opalina*, found in the faecal matter of the frog's rectum, are examples. The protozoa which occur in the guts of termites, and the bacteria which colonize the rabbit's caecum, make possible the utilization by the host of cellulose derivatives, and are symbionts.

Parasitism of a peculiar kind characterises the European cuckoo, *Cuculus canorus*, whose habits include a breeding pattern which results in the young bird being reared by a different species and expelling the eggs or nestlings of its fosterers in the process, thereby securing for itself the exclusive care of the adults. The relationship between the young cuckoo and its fosterers would be commensal were it not for the expulsive behaviour. This is triggered by a thigmotactic response to contacts between solid objects, such as eggs or nestlings, and a sensitive hollow in the back of the unfledged cuckoo. The eggs or young birds are rolled into this hollow and tipped out of the nest by the blind cuckoo which obviously cannot see what it is doing. The expulsive instinct has disappeared by four days, long before which time the parasite usually has the nest to itself (*Handbook of British birds*[84]).

As it is, this species of cuckoo has become so well adapted for its particular brand of parasitism as sometimes to succeed in causing other birds, not in possession of the nest wherein it is lodging, to feed it and, in so doing, to reduce the quantity of food delivered to their own young. Darlington has filmed one individual cuckoo which was tended by a pair of meadow-pipits in whose nest it was being reared and, shortly after becoming the sole occupant, also by two pied wagtails which, at the time, had three unfledged offspring. Only one of these left their nest successfully. Here, one of the stimuli was apparently vocal, the wheezy note of the fledgling cuckoo being both persistent and penetrating. The rearing of this young cuckoo was certainly the cause of the death of five birds (the meadow-pipits ejected as eggs) and possibly also of the two unfledged wagtails.

An advantage of the parasitic habit in this particular species of cuckoo might originate from the high proportion of hairy caterpillars in the diet of the adult. Although digestible by a grown cuckoo, they could be fatal to a nestling. Birds which become fosterers gather smooth invertebrates and normally reject hairy ones.

Symbiosis
Symbiosis can be regarded as a relationship in which two species live in a close association from which each derives some benefit. The association is not necessarily obligatory. It may arise as an incidental development from parasitism. The nitrogen-fixing bacterium *Rhizobium* may be cited as an example.

Rhizobium
Rhizobium occurs free in soils but also parasitizes the roots of

papilionaceous crops and other plants, which it invades through the root-hairs and where it causes the formation of galls with localized proliferation of the host's tissues, a common reaction of plants to some kinds of parasitic attack. The holophytic host synthesizes monosaccharides in excess of its metabolic needs, thereby benefitting *Rhizobium* by supplying carbon, hydrogen and oxygen from part of which, together with atmospheric nitrogen, the bacterium brings about the formation of amines: such compounds, in turn, are excessive to the invader's own requirements and one stage in protein synthesis by the host is supplemented or even circumvented.

On its own, the host can achieve full protein metabolism; but the benefit to it of the bacterium is measurable under experimental conditions, when comparisons between crops grown on sterile pabula and those infested with the bacterial nodules show the latter to grow more quickly and, in many cases, to acquire at maturity a greater biomass.

Similarly, the mycelial fungi which infest roots as mycorrhizas, functioning as root-hairs and fixing nitrogen, probably evolved as symbionts from parasites.

Advantage to the larger partner is observable in green hydra (*Chlorohydra viridissima*) where the alga *Zoöchlorella* colonizes cells of the endoderm. Specimens cultured in well illuminated situations produce more buds asexually, and come into sexual phase later, than do those reared where the illumination is low, provided that both samples have access to similar quantities of prey in the form of small crustacea. It could be that the elimination of ammonia, carbon dioxide and salts from the body is difficult for coelenterates and that this, in part, explains the occurrence of unicellular algae in the endoderm of so many. There is need for further experimental work here.

In lichens, the symbiosis appears to be an enforced condition arising from the resistance of algae (assimilatory) to attacks by fungi (non-assimilatory) so that both organisms survive and grow in association together and, in fact, are able to colonize places where neither on its own could persist for long. Although lichens are commonly regarded as specific entities with symbionts which are interdependent, forms occur showing such gradation as to make classification difficult.

Thus, some of the crustose lichens established as coatings over solid objects may exhibit extremes on the bark of the same tree trunk and correlated with ecological variants which are more or less clearly defined: growths range from algal cells in the wettest parts free from an admixture of fungal mycelia, through the intermediate condition of lichenized algae, where relatively large clumps of algae lie among small and isolated wefts of hyphae, to an increasing prevalence in the driest places of the fungal component. Despite the customary acceptance of a lichen as exemplifying typical symbiosis, the two partners are demonstrably unequal in benefitting from the association: lichen fungi (mainly Ascomycetes) are known as free-living organisms only in laboratory cultures on sterilized agar-media. In contrast, the algae which occur within lichen thalli are less specialized and include species which are found living free, often in areas adjacent to those supporting the lichens. In this connection, it is noteworthy that lichens have developed on a large scale dispersive bodies, such as ascospores and pycnidiospores, which comprise only the fungal component, from which lichenized situations can arise if the propagules happen to reach places where the appropriate algae already exist.

But certain lichens have acquired other methods of reproduction where the propagule carries both associates, fungus and alga. Some release powdery granules (soredia): under the microscope a soredial granule is seen to consist of a little tangle of fungal hyphae binding together a few algal cells. Soredia readily become detached and wind-borne. Isidia are structures larger than soredia into which a brittle thallus readily fragments and which, under favourable conditions, may set up lichen thalli elsewhere.

On a few lichens, wart-like outgrowths, cephalodia, contain blue-green algae despite the fact that the assimilatory partner in the thallus is a green alga. The nitrogen-fixing properties of the blue-greens may be of special benefit to the fungus; and it is noteworthy that certain lichens (e.g. *Peltigera*) having blue-greens as ordinary thallus algae, do not develop cephalodia. (See also Duncan[19], and Kershaw & Alvin[38]).

From the foregoing, the fungus appears as the more dependent of the two associates, which is in keeping with the supposition that a lichen represents parasitism of a fungus upon an alga. But one advantage derived by the alga from the mycelium could be that establishment is facilitated in situations unfavourable for colonization by the alga alone, the hyphae retaining moisture where conditions otherwise are those of aridity.

Relationships between associated organisms which are mutually advantageous and arise from those which initially were one-sided in their benefits may be so phased with the life-histories of the partners as not to develop simultaneously. Freshwater mussels and bitterlings provide an example.

At the onset of breeding, the female of the European bitterling-minnow, *Rhodeus sericeus*, develops from the vent an ovipositor deeper than her body, the tip of which she places in the mantle-cavity of a freshwater mussel through the lamellae of the inhalant siphon. Accuracy of the insertion requires a special degree of adaptation in that the fish cannot see what she is doing, and the siphon is particularly sensitive to contact and stimulates the severe clamping of which the mussel is capable and which, if applied, might be expected to crush or sever the duct. Nevertheless, she deposits her eggs a few at a time in a number of mussels: immediately after oviposition, the male fish sheds sperm into the water which passes into the mantle-cavity with the inflow current and effects fertilization of the eggs. Here they undergo development in protected conditions. After absorption of the yolk-sac the fry feed as commensals for a while on suspended matter drawn in as food by the mussel; but should the bivalve's own spermatozoa be undergoing release at the time, they may take these when parasitism, albeit incidentally, becomes an additional feature. Eventually the young fish make their way into open water through the exhalant siphon.

When the bivalve itself reproduces, its eggs develop into glochidial larvae within cover of the parental structures: subsequently the glochidia ejected in the exhalant flow, are free-swimming for a short period, and become attached by an adhesive byssus-thread to the bodies of passing fishes of various species, setting up irritation with external blistering by the marginal teeth on their valves. The larvae feed parasitically on tissue-fluids inside the superficial swellings, and metamorphose into small mussels before the blisters rupture and release them from their hosts to fall into the bottom mud where they continue as free-living animals. Infested fishes usually recover.

Adults of the larger freshwater bivalves, *Unio* and *Anodonta*, are practically sedentary. Their maximum speed has been estimated as between 500 m and 1 000 m a year from Darlington's observations, made during the two years 1935–7, of marked specimens of *Anodonta* living on the bed of the Lancaster and Kendal Canal. But this is not the distance they effectively covered: indeed there was no evidence that significant progression in any one direction ever occurred. Clearly, temporary parasitism of far-ranging animals combines the advantages of a high-protein diet during metamorphosis with dispersal.

When bitterlings become the hosts of the glochidial larvae the reciprocation, viewed as a whole, can be regarded as that of facultative symbiosis.

Commensalism

Commensalism, a looser relationship than symbiosis, is often defined as one based on the sharing of food, in which there is generally a primary organism in some way supplying the food, and from which all the members usually benefit.

Such a definition applies fully to the association between zic-zac birds and crocodiles, where food particles are picked from the predators' teeth by the smaller animals. It is partly true of the link between man on the one part and the house mouse and house sparrow on the other. Here the relationship is one-sided since man derives insignificant benefit from it whereas the mouse and sparrow are so dependent on food provided by him that, were he to disappear they, too, would vanish. This happened in the case of the St. Kilda house mouse (*Mus musculus muralis*) following the abandonment of the island in September, 1930. The house sparrow, being dependent basically on grain, is a commensal of agricultural man. Possibly it depends also to some extent on his provision of certain kinds of shelter. Rooks are commensals in being largely dependent, directly or indirectly, on man's agricultural practices for most of their food. Paradoxically they establish the majority of their nests in his settlements, notwithstanding the fact that he is their principal enemy. More tenuous still is the temporary commensalism between the sheep and such birds as starlings which sometimes take ectoparasites from its fleece but have numerous alternatives in the way of food; and much the same is true of barnacles (*Coronula* sp.) attached to whales, the gain for these sessile filter-feeders is that of transportation.

Commensalism, then, covers the provision by some living organisms of any of the basic habitat needs, including shelter, of others. It is a matter degree and a relationship passing imperceptibly into that of the usual food chain connections enlarged upon in

Chapter 5. Such merging, where the primary agent becomes peripheral in some of the influences it exerts, is seen in the complex which develops at the sites where the Florida gopher-tortoise is active. The tortoise excavates deep holes in sand and this material, broken up initially by the tortoise, is utilized for burrowing by another reptile, the red-tailed skink. Flat beetles colonize the depressions and eat the tortoise dung while cave crickets, so specialized for a cryptozoic mode of life as to be nearly blind, devour the beetles' excreta as well as the mycelia of fungi growing in the organic-rich walls of the excavations. Flies, attracted to the situation, are taken by spiders in webs spanning the entrances to the burrows as well as by gopher frogs waiting for them just below the entrance.

Clearly, distinctions like those outlined above, are arbitrary and indicate points along a continuous sequence, ranging from the extreme of parasitic predation where the death of one associate inevitably supervenes, to situations in which the partners are linked by mere gossamer threads.

5 Energetics

5.1 Origin of the energy flow

To state simply that the energy any organism needs for its metabolism is derived from the breakdown of its food during respiration, only begs the question of what was the source of the energy in the first place. In ecosystems generally, the effective supply of energy is that radiated by the sun and, during photosynthesis, used by chlorophyllous plants for bonding into large molecules the elements obtained from the small molecules of carbon dioxide and water. Direct utilization of energy in this way is a feature of autotrophs. Among the autotrophic organisms additional to green plants are those bacteria which obtain their energy, not from the sun, but by the breakdown of comparatively simple inorganic molecules in the surroundings. The energy thus released is used by the bacteria for the bonding process of chemosynthesis, somewhat after the manner of photosynthesis.

5.2 Trophic relationships within the ecosystem

Autotrophs are the producers. Upon them the rest of the living world, the heterotrophs, depend for their food, either directly or indirectly. Heterotrophs are the consumers. An ecosystem must contain both producers and consumers and the trophic relationships between them involve some of the most important interactions within a community.

Autotrophic green plants are the primary producers and make up the first trophic level: they are eaten by herbivores, the primary consumers; these in turn are taken by carnivores, the secondary consumers; and a fourth trophic level is common, that of the larger carnivores, the tertiary consumers. These are the principal links in a food-chain or food-web.

In a complex situation, those organisms which obtain their food from the producer plants by the same number of stages are said to belong to the same *trophic level*. Such a classification is one of function and not of species, for a particular species may occupy more than one trophic level within a community, according to the kinds of food eaten. Man is a good example, for he derives energy by eating foods of both plant and animal origin.

5.3 Food chains and food webs

Within any ecosystem, large or small, it is possible to recognise a number of food chains. An oft-quoted example is that of the open sea. Here the phytoplankton, principally composed of the diatom *Chaetocerus decipiens* and some flagellates, are used as food by many animals. One is a common copepod, *Calanus finmarchicus*, which forms the diet of the adult herring, *Clupea harengus*. This series forms a simple linear food chain:

C. decipiens→ *C. finmarchicus*→ *C. harengus*

to which might be added a secondary consumer, the cormorant, *Phalacrocorax carbo*.

Another simple food-chain noted by Leadley Brown in the Galapagos islands again had its origins in plant life. The food of the marine iguana, *Amblyrhincus cristatus*, is kelp found in the shallow, well-illuminated waters round the archipelago. On occasion, the Galapagos hawk, *Buteo galapagoensis*, was seen to attack and devour the young of *A. cristatus*. Another food chain, the last three links of which were clearly witnessed on Tower Island, was:

Phytoplankton→ zooplankton→ fish→

Galapagos storm petrel→ Galapagos short-eared owl
(*Oceanodroma tethys*)   (*Asio flammeus galapagoensis*)

Food chains on volcanic islands of recent origin, tend to have their first links in the sea where there is an abundance of phytoplankton compared with the paucity of terrestrial vegetation.

In the tidal waters around the British Isles, the barnacle, *Balanus balanoides*, is preyed upon by the dog whelk, *Nucella lapillus*, which uses its radula for rasping apart the plates of the barnacle. However, it takes longer to do this than actually to consume the contents so, by

Fig. 5.1 A FOOD WEB ON THE GALAPAGOS ISLANDS.

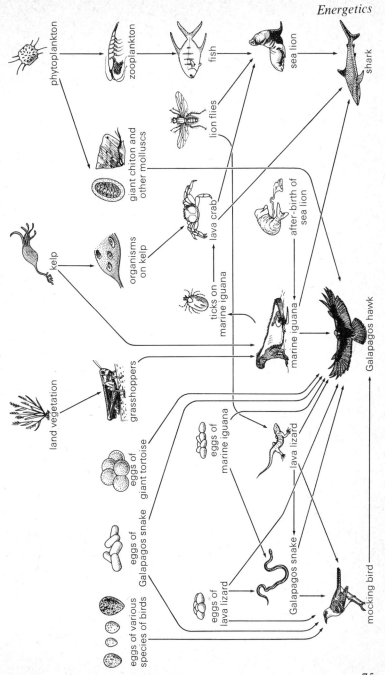

concentrating on the larger specimens of its prey, *Nucella* increases not only its own production but that of the barnacle, whose population is often limited by overcrowding.

An interesting variant is shown by separate faunas often occupying the crevices in the spray zone of a rocky shore, both include scavengers, herbivores and their predators and both exhibit vertical and horizontal zonation – the vertical zones resulting from the time each part of the area is submerged by the sea and the horizontal zones dependent upon the distance from the mouth of the crevice to its limit. Collembola feed on detritus among the lichens encrusting the rock, while mites and fly larvae are the predators. These are the terrestrial fauna operating when the tide is out. As the tide and spray cover the crevice, polychaete worms move in as the main predators.

It is clear that a simple food-chain rarely exists as such, for by definition it implies that each species represented in the chain feeds on a single species. This is the exception rather than the rule, for the food of an organism is more likely to comprise several different species. These in turn may eat a variety of species. Thus if all the food relationships could be worked out, a simple chain would soon become a very complex web. The trouble lies in the temptation to oversimplify, with the result that the diagram of the food web of a community which is finally produced does not convey a correct picture of the interrelationships existing.

The simple food-chain quoted above for the Galapagos hawk could be extended to the web illustrated in Fig. 5.1. This shows some of the links of the web which were either observed directly or are documented elsewhere. Doubtless it would become even more complex were all the links known.

From the diagram several important generalizations of food-webs can be made. The hawk and the shark are clearly the terminal predators, but the hawk feeds on a variety of prey including species from the carnivore and herbivore trophic levels. The shark preys on sea lions and their young and doubtless many other marine vertebrates not shown in the diagram as well as upon young and adult marine iguanas. During its development, different stages of a species

Fig. 5.2 FOOD-WEB OF SHORT-EARED OWL. Some of the links of this terminal predator when it frequents coastal sandhills. Note its dependence on both terrestrial and marine situations. Blakeney, Norfolk.

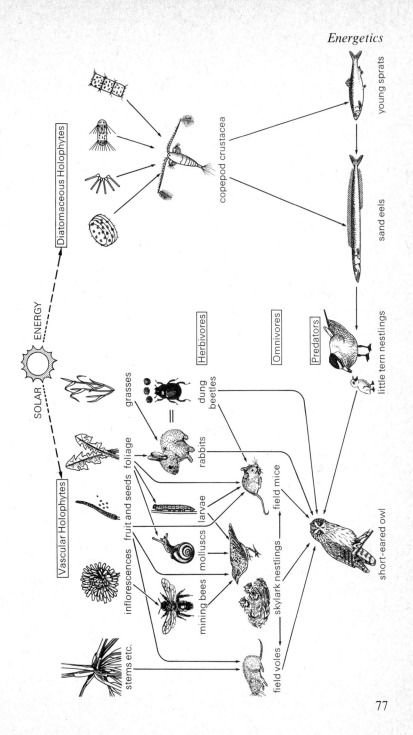

77

feed on different foods and are attacked by different predators. The iguana is a case in point, for the young tend to feed on intertidal rock-encrusting weeds and lichens while the adults forage to greater depths for *Sargassum* weed. It is interesting that, although supposed by earlier explorers of the Galapagos to feed exclusively on kelp, Eibl-Eibesfeldt[22] observed several cases of adult iguanas devouring the afterbirths of sea lions and also in one case, taking a grasshopper. Karl Angermeyer, a settler on Santa Cruz, one of the central islands of the archipelago, has a colony of tame iguanas which daily visit his dwelling for household scraps.

The ticks living parasitically under the scales of the iguana are eaten by the lava crabs which are often to be seen perched on the back of an iguana searching for them. The mocking bird can be considered as a predator of many animals in the community, for besides feeding independently upon lion flies and other insects, it will always be found in the vicinity of the nests of both birds and reptiles and makes a quick attack on their eggs if the nests are left unguarded.

Again, the food-web of the short-eared owl, shown in Fig. 5.2, gives only the main organisms involved at each trophic level and no attempt is made to name the numerous species of autotrophs consumed by the herbivores. It might well be that further investigation would reveal the presence of other animal remains, such as those of the mole, and other small birds in the food pellets.

Nevertheless the diagrams indicate the chief food preferences of these organisms and such knowledge can become of economic importance especially if food-webs include the food of man, as is the case with the classic web of the herring worked out by Hardy[31]. Here a knowledge of the lower levels of the web – the phyto- and zooplankton – may be used to predict the likelihood of the presence of the top predators in a habitat. Such knowledge may be of great assistance to fishing trawlers in locating shoals of herring.

It is evident that food-webs have two distinct disadvantages as sources of information concerning feeding relationships. They must either be so complete involving the differentiation of hundreds of species that a diagram would become impossibly complex, or so oversimplified as to be virtually useless.

5.4 Ecological pyramids

'Who eats whom?' has been a subject of interest to natural historians

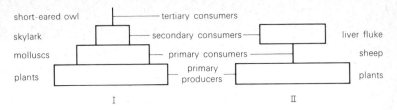

Fig. 5.3 PYRAMID OF NUMBERS: I Where the primary producers are small in size and large in numbers.

II Where the primary consumers are large in size and small in numbers and are preyed upon by a large number of parasites — the secondary consumers.

for more than a century. But it was not until 1927 that Elton[23] pointed out that the number of links in a food chain between primary producer and final consumer are not usually more than three. However, he also drew attention to the obvious disparities existing between the numbers of organisms at producer level and the numbers at the top level. A typical community is composed of a large number of green plants at the producer level, while the herbivores and carnivores constitute successive levels, rapidly decreasing in numbers with each level until at the top there are small numbers of larger predators. Such a succession is a 'pyramid of numbers' (*Eltonian Pyramid*) where there is a progressive decrease of numbers with a corresponding increase in size of the organisms constituting the pyramid.

Based on numbers only, a pyramid such as that in Fig. 5.3 is the one commonly represented in text books. It omits, however, an important group of organisms which, almost without exception, are present in every community. These are the decomposers, the fungi, bacteria and other detritivores, always associated with decay and the recycling of matter. If such decomposers, were represented, there would be a broad upper level added above the tertiary consumers.

In nearly all communities one or more of the organisms forming the different trophic levels will support a number of parasites. The simple pyramid of numbers illustrated in Fig. 5.3 becomes inverted at the secondary consumer level for one sheep can be infested by large numbers of fluke.

5.5 *Energy flow through an ecosystem*

In any ecosystem, an attempt to quantify the information supplied by food-webs requires a knowledge of the rate at which food is assimilated as well as the respiratory rates at each trophic level: i.e. the amount of food material retained at any one moment (the biomass) and the rate at which it is metabolised. Here again, it will be realised how great are the disadvantages of using *numbers* of organisms required to support a given number of other organisms, for this leaves one with the impossible situation of comparing widely different organisms, some of great size, with those which may be microscopic. By using biomass as a comparison, the whole situation assumes validity and two communities can then be compared on a quantitative basis.

Whereas heat is gained by and lost to an ecosystem, the transfer of energy between two successive trophic levels obeys the first law of thermodynamics, namely that it can be transformed from one form to another but is neither created nor destroyed. The ecologist is concerned with what happens to the energy once it has entered the ecosystem and before it leaves it.

The first stage of the process of energy flow is the transformation of solar energy into chemical energy which is brought about by the photosynthetic activities of green plants. The amount of energy transformed by these autotrophs is a small fraction, about 1–5 per cent, of that available to them. The rest is lost as heat. These primary producers store the chemical energy in their tissues and the amount stored per unit area per unit time is their *gross primary production*. Some of this stored energy will be released in respiration. What is left is the *net primary production* and this is the amount available to the next trophic level. Phillipson[60] estimated the net primary production to be about 80–90 per cent of the gross but it has been pointed out by Macfadyen[47] that this ratio can vary in different climates being as low as 25 per cent in tropical rain forests.

One of two things can happen to the energy-rich organic material thus stored by plants.

1 It may be eaten by herbivores forming the next trophic level where some will be liberated during respiration, some will be excreted as faeces and some will be assimilated as body protein. This latter store of energy forms the net productivity of the herbivore level.

2   When the vegetation dies back it will pass into the soil ecosystem as litter. Obviously the flow of energy into these two systems will vary in proportion from one area to another and be influenced by man where agricultural practices may alter the structures of communities.

The energy stored at the herbivore level is available to the next trophic level, the carnivores, which in turn lose some of the energy taken in as metabolic heat, retaining some as stored energy for the top carnivore level.

This is an oversimplified picture, for in reality where complex food-webs, and not simple chains, are concerned, energy does not flow directly from one level to another but meanders and recirculates within the various trophic levels.

If, therefore, a *pyramid of energy* is constructed instead of a pyramid of numbers, we arrive at a quantitative representation of the total amount of energy retained, exported and lost by metabolism at each trophic level.

Energy pyramids constructed for different ecosystems vary enormously. Such factors as the amount of incident light falling on the plants may result in a larger figure for energy retained at the producer level in a grassland ecosystem than in a shaded woodland. Seasonal differences of light intensity and temperature may also affect the situation. In deciduous woodland development of a superior leaf canopy will intercept the light arriving at the herb layer so that different figures will be obtained for this ecosystem at different times of the year.

Again, age and sex of the populations within a community will affect the amounts of energy retained and transported. O'Connor[55] and Phillipson and Watson[61] found that at 20° C, a population of young Enchytraeidae are metabolically more active than older ones and that in the woodlouse, *Oniscus asellus*, variation in respiratory activity in the adults was related to variations in reproductive activity associated with gonad development. In a soil ecosystem the generally low metabolic rate of a community of woodlice is compensated by a high biomass in certain areas. Conversely the low biomass of Collembola, Nematoda, and Enchytraeidae can be offset by their relatively high metabolic rates.

In nearly all ecosystems there are two types of energy chains operating. The grazing chain is one in which the primary producers are consumed by animals, whether they be large such as cattle or small such as zooplankton. These are the herbivores in which group we can also include plant parasites. The rest of the net primary production is destined to be consumed as dead material of one kind or another. Such things as dead leaves falling into a pond as well as other dead plant material which is reduced to detritus by various important groups of *detritivores*. Thus we have a *detritus chain* involving the flow of energy resulting from the breakdown of dead matter.

To the grazing and detritivore chains must be added another where parasites of the herbivores and carnivores are concerned. Some authors prefer to include parasites as part of the same trophic level as their host and yet this is not strictly true in all cases since the energy stored by the parasite at the herbivore level can be either transferred to the detritus chain on the death of the parasite or persist after death of the host should the parasite itself be preyed upon either by a hyper-parasitic species or by another predator. A separate *parasitic chain* therefore exists.

5.6 Rate of energy transfer

It was due to the American ecologist, Lindemann[44] that interest became centred on the measurement of the rate of transfer of energy through the different trophic levels of an ecosystem. This lent meaning to the new concept of community dynamics and involved new techniques for measuring energy input, energy content and loss at each trophic level and in some specific investigations, of these parameters at species level.

Since, as we have already said, transfer of energy from one trophic level to another follows the laws of thermodynamics, it can never be 100 per cent efficient, for at each level there is a progressive loss of energy in the form of heat. Nevertheless the solar energy reaching a system is equal to the heat energy leaving that system.

The interpretation, in simplest terms, of Lindemann's concepts of community dynamics is illustrated in Fig. 5.4 where the Greek letter lambda (λ) signifies the energy content (energy retained) of each trophic level. Thus λ_1 represents the energy content of the producers, λ_2 that of the herbivores (or detritivores) and so on. R represents the loss through metabolism of heat energy per unit time.

R_1 being that lost by the producers, R_2 by the herbivores and so on. The energy budget of a population can be represented by the simple equation $I = R + \lambda$ where I is the energy assimilated with the food, R that lost through metabolism and λ the energy retained (the production) and therefore available to organisms of higher trophic levels. The value of I, or the difference between the amount of food ingested and that egested per unit time per unit body weight can be obtained by subtracting the weight of faeces produced from the weight of food consumed. The amount of production (λ) can be estimated by making successive measurements of biomass. This can create considerable difficulties where micro-organisms are concerned and indirect methods have often to be resorted to. The values obtained, by whatever means, for biomass and assimilation can be converted into calorific equivalents by burning the proto- plasm and then the food material in a bomb calorimeter; in the case

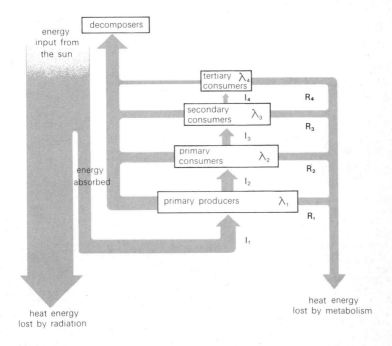

Fig. 5.4 SCHEMATIC REPRESENTATION OF ENERGY FLOW THROUGH AN ECOSYSTEM. Based on Lindemann's concept of community dynamics. Lindemann[44].

of micro-organisms, various forms of microbomb calorimeters have been developed. The factor R can be estimated directly by measuring oxygen uptake. Again, in the case of micro-organisms indirect methods may have to be used.

Because many factors cannot easily be investigated under natural conditions, it is obvious that there are technical difficulties involved in making any accurate assessment of how the stored energy is shared between the groups. The situation is further complicated by fluctuations in biomass and density which occur under natural conditions and which therefore influence the total metabolism of a

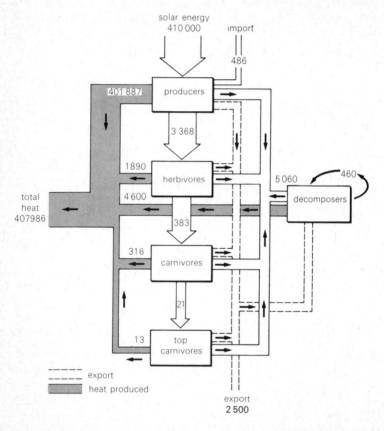

Fig. 5.5 SCHEMATIC DRAWING OF THE FLOW OF ENERGY IN SILVER SPRINGS, FLORIDA. Figures in kcal/m year. (Adapted from Odum[56]).

system. Not only this but rates of feeding, assimilation and metabolism vary according to the biotic, physical and chemical conditions prevailing at the time. For instance Newell[54] investigating
the oxygen consumption of three species of slug over a 24-hour
period found that they went through a quiescent and active period
which profoundly altered their oxygen consumption. By determining the duration of their quiescent and active phases and the oxygen
consumed during these periods over 24 hours, the actual mean rate
of consumption of oxygen per day was considerably lower than if a
constant respiratory rate had been assumed. Again the metabolic
rate of a poikilothermous species varies with the temperature and
stage of development, sex and breeding conditions as already noted.

It is not surprising, therefore, that work of this kind is in its
infancy, Progress is dependent upon perfecting means of measuring
energy losses and gains in a population under natural conditions.

One such study was made by Odum[56] who investigated a spring
system in Florida. Here Odum found an abundance of decomposers
feeding on food produced within the ecosystem. The importance of
these decomposers in this ecosystem is shown by Odum's results
(Fig. 5.5) where 5 060 kCal/m²/yr flowed through the decomposers
while only 3 368 kCal/m²/yr went through the grazers (1 kCal=
4.1868 kilojoules).

5.7 Ecological efficiency

Fig. 5.5 shows that most of the radiant energy available to an
ecosystem is never incorporated, while a great deal of that incorporated is dissipated at each trophic level in the form of heat. But
how efficiently is energy transferred from one level to another? In
any ecosystem the amount of energy made available to the next
trophic level divided by the amount of food ingested by the prey,
gives a measure of the *ecological efficiency* of the energy transference
between those levels. This can be represented thus:

$$\text{ecological efficiency} = \frac{\text{energy contained in yield to predator}}{\text{energy content of food ingested by prey}}$$

and the sum of the efficiency of transference of energy throughout
the trophic levels is the *gross ecological efficiency* of the ecosystem.

Theoretically, maximum efficiency will prevail if there are

sufficient organisms at each trophic level to consume during the year, the organisms of the level below, down to the level where they are never numerous enough to be short of food nor depleted in numbers to the point where there is a food surplus. In other words where the food of the prey is being exported to the best advantage.

6 Stable situations

Somewhat uncritically, perhaps, colonization of a site is envisaged as a protracted series of events, during which a succession of communities is built up, each providing the conditions favourable for the one which follows. Such communities are held to be unstable and to undergo changes both in the composition of their components and the population sizes of these species. Frequently, conspicuous and comparatively rapid succession is apparent in those freshwater situations like disused canals and neglected ornamental lakes on former estates (Fig. 6.1), where silting up gives rise, through reed-swamp, to terrestrial forms. Finally, successional changes are considered to peter out in the establishment of a stable community.

Fig. 6.1 SUCCESSION THROUGH SILTING-UP. Reed-swamp vegetation accumulating mud, choking the channel which used to catch open water supplying a fishing lake, and gradually invading the lake itself. Terrestrial communities are extending as the ground becomes drier and new soil piles up. Malvern College Nature Reserve, Herefordshire.

6.1 What is a stable community?

Although commonly regarded as the automatic climax of succession where nothing happens to arrest the process at some intermediate stage in development, the stable community, taken literally, is as much a myth as infinity. There is the traditional view that stability is reached and then maintained through the prevailing physico-chemical and climatic conditions, when the community becomes dominated by the largest plants able to exist under these conditions, when further vegetational changes are those of replacement of pre-existing species by equal numbers of forms identical with them, and when the animal constituents are presumed to have attained comparable stability. Such a conception is an ideal not found in practice since no environment is stable. Even by its death and collapse, a single tree in a woodland climatic climax has a decisive effect on many of the underlying subordinates until such time as its place has become occupied by another individual of the same kind and stability thereby restored.

If animal communities are considered, factors may appear which cannot be compared with anything found in plants. Thus, 4–5 Sika deer released on James Island, Chesapeake Bay, in 1916, had built up to a population of some 300 by 1956; but half the deer died in the first three months of 1958 and the population levelled off at 80 in the following year. Christian in Hall S.[30] investigating this community, found that the population density was one deer to the acre and he considered the food supply to be adequate; but he found abnormal cell structures in the adrenal glands of those animals dying during the crash and concluded that they had perished not from shortage of food, but from stress.

That alterations can involve even the dominants in a 'stable' climax appears from recent developments affecting certain oak woods of long establishment in the east Midlands and elsewhere: here the sycamore, *Acer pseudoplatanus*, is moving into spaces left by the death of oak trees and increasingly taking them over. Sycamore, an alien introduced in the middle of the fifteenth century from the mountainous parts of Europe, has now reached such high populations in localities where this happens as to be at an advantage in competition with the indigenous tree. Its numerous seeds are effectually dispersed in wind-borne fruits—the writer has also found damaged sycamore fruits stored by mice—in contrast to the heavier acorns of oak, and germinate into young plants capable of tolerating the somewhat dappled form of illumination imparted by the oak canopy beneath which they grow rapidly. Sycamore canopy is dense and exercises marked effects on the subordinates, killing off the slow-growing seedling oaks and influencing many species by changes in lighting, and also by changes in the pattern of precipitation brought about by its leaf mosaic, which inhibits the widespread scattering of small droplets passing through the oak foliage but retains the water so that it falls repeatedly in large drops on the same localized parts of the woodland floor. Sycamore is now occupying some of the gaps left when native trees are killed by Dutch elm-disease. Notwithstanding its attractive appearance when growing alone and its usefulness for timber, it can become an aggressive weed.

When beech, *Fagus sylvatica*, a native with a dense canopy, develops in sufficient numbers near to a stable oakwood climax and sufficiently close to it for interspecific competition to supervene, the beech frequently becomes the 'winner' and the oak the 'loser'.

As a rule, a climatic climax incorporates numerous habitats, many of them more or less in a state of flux: as a concept it is convenient in emphasizing the slowing down of changes–particularly those originating within the ecosystem–until they pass largely unnoticed. The longevity of the dominants is one reason for the changes becoming less apparent. The whole climax concept is largely dependent on choosing the right time scale, anyway.

6.2 Palaeo-ecology and past successional changes

Several workers have reconstructed long-term vegetational changes developing in postglacial times by analysing the pollen-content of successive layers of peat. From evidence of this kind have been deduced changes in so-called stable communities and their correlation with changes in the climate. Where pollen is to be examined the work calls for special skills in extraction and identification; but Darlington has utilised animal fossils, which are larger and easier to extract, examine and identify, to establish some of the changes likely to have occurred from the beginning of the Ice Age, in a coastal region of Essex and Suffolk. The region includes the sites of present-day Aldeburgh and Walton-on-the-Naze.

Following a long period of steady subsidence beneath a calm sea when the chalk, characteristic of much of lowland Britain, was laid down, the whole area of East Anglia southward to the valley of the present Thames was raised as a land-mass and subjected to tilting and folding. Subsequently, part of the region was again invaded by a sea which now deposited Eocene beds, mainly clays and sands. The Eocene formation exposed on the coast here is London Clay, a slippery blue-grey deposit weathering to brown flakes and lying at the base of the low, sandy cliffs which are the particular scene of the present section. After an interval of time, towards the end of which the Pliocene period was succeeded by the onset of glacial conditions and the dawn of the Pleistocene period, the region became submerged beneath a precursor of the North Sea. Movements of the floor of this sea controlled the distribution of later deposits – the shelly sands and pebbly gravels called *crags*.

At the Naze, the formation is Red Crag, a shore-deposit of current-bedded sand, with bands of laminated ironstone and bearing strata very rich in fossil gastropod and lamellibranch shells. These remains, often broken and abraded in the older levels of the crag,

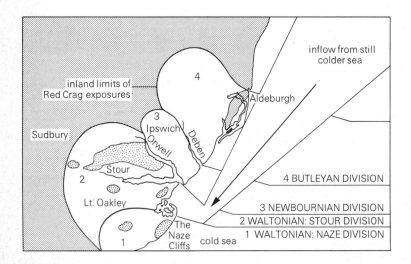

Fig. 6.2 PALAEO-ECOLOGY. Hypothetical Pleistocene bays near the borders of present-day Essex and Suffolk deduced from animal remains in Red Crag strata. The supposed sequence of formation is indicated numerically.

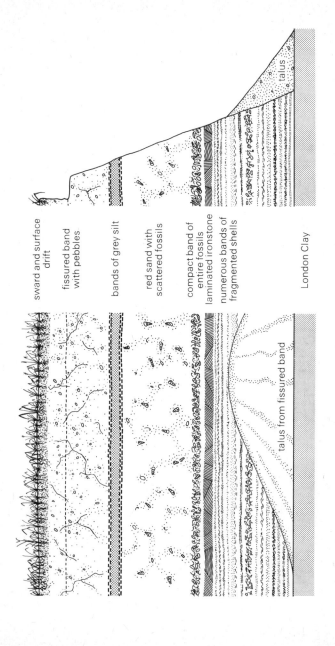

sward and surface drift

fissured band with pebbles

bands of grey silt

red sand with scattered fossils

compact band of entire fossils
laminated ironstone
numerous bands of fragmented shells

London Clay

talus from fissured band

talus

talus

Fig. 6.3 PALAEO-ECOLOGY. Red Crag strata in the cliffs at Walton-on-the-Naze: *left*, frontal aspect; *right*, in profile. Stormy conditions during the earlier stages of formation are suggested by smashed-up shells in the lower levels. Height of exposure above London Clay, 6 m.

indicate that the material was piled up near the shores of land-locked bays by easterly gales prevailing at the time. In addition to the enormous number of molluscan fragments, the Red Crag yields foraminifera, corals, sponges, branchiopods, polyzoans, crinoids, echinoids and barnacles. Originally referred to the end of the Pliocene period, Red Crag is now regarded as early Pleistocene and its formation probably coincided with the First Glaciation. Its maximum thickness is 13 m and much of it is covered with glacial drift. Pollen grains of dwarf willows and birches–shrubs of an arctic-alpine flora–have been recovered from this drift. Apart from its elevation into cliffs, other changes since its formation include the removal by percolation of calcium carbonate from the upper levels, with the accumulation of limy material in some of the lower fissures, and deep red staining from the oxidation of ferrous compounds.

Although subdivisions of geological deposits usually succeed vertically, here, as a result of earth movements, the shores on which the beds were laid down gradually shifted northwards, with the result that those in the Red Crag succeed horizontally from south to north without overlapping. In this direction, the fossils occurring in them show a greater percentage of present-day and northern forms, and fewer extinct and southern species or non-marine animals. As the bays shifted towards modern Aldeburgh, cold water apparently entered from the north, killing off many of the southern molluscs and bringing with it an increasing number of northern types. Even in the few outcrops of the crag south of the R. Stour, the fauna in the Little Oakley deposit, only five miles to the north-west of Walton-on-the-Naze, is distinctly more modern than that in the Walton cliffs: it was here that a few Arctic molluscs made their furthest penetration southwards.

Four stages of Red Crag can be recognised: Waltonian–Naze division, the most southerly and oldest; Waltonian–Stour division and Newbournian, the intermediates; and Butleyan, the most northerly and recent. Darlington's reconstruction of the hypothetical Pleistocene bays founded on fossil evidence is illustrated in

Fig. 6.4 PALAEO-ECOLOGY. Eleven common molluscs from the Waltonian Red Crag. Six are extinct. The majority, including the extinct species, had non-pelagic larvae, which suggests a cold sea low in plankton (cf Orkney today). Note the swollen whorls of the extinct gastropods wherein the sedentary young were supposedly retained. *Scale*, line represents 1 cm.

Extinct species

Nucella tetragona

Hinia reticosa

Neptunea contraria

Spisula arctuata

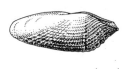

Barnea (Pholas) cylindrica

Corbulomya complanata

Extant species

Nucella lapillus

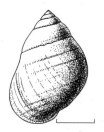

Littorina littorea

Turritella communis

Dosinia exoleta

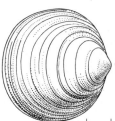

Glycymeris glycymeris

93

Fig. 6.2, and Fig. 6.3 gives the present appearance of the middle part of the Naze cliffs. Examination of these cliffs reveals the many strata of shell fragments in the lower (older) levels, suggestive of fracture in stormy conditions during the early stages of formation of Red Crag; and the strata bearing undamaged fossils in the upper deposits, which were evidently laid down when more tranquil conditions had set in.

A small selection of the commonest molluscan remains obtained from the cliffs here is set out in Fig. 6.4. A particularly interesting species is *Neptunea contraria*, a whelk-like shell with a left-handed spiral instead of the usual twist to the right.

Such a palaeo-ecological investigation requires nothing in the way of elaborate equipment but it has wide application for other regions and situations. It is clearly within the capabilities of sixth-form students desirous of undertaking project work. For students, the various British Regional Geology Handbooks, prepared by the Geological Museum for its Survey, are especially useful publications.

6.3 Climax and subclimax ecosystems

It is widely accepted that, in Britain, climatic climaxes of primeval origin are infrequent and largely restricted to somewhat spectacular situations. Examples include rocky shores, mountainous uplands, and certain timbered regions in Scotland harbouring elements of a supposedly ancient fauna: the pine-tracts of the Spey valley with the Scottish subspecies of the wild cat, *Felis silvestris grampia*, and the Scottish form of the crested tit, *Parus cristatus scoticus*; and the birch-clad slopes in Perthshire which retain a magnificent moth in the Rannoch sprawler, *Brachionycha nubeculosa*.

Three assumptions are common:

1   that the natural climax over lowland Britain is oak forest;
2   that 'natural' forest is only represented by occasional relics such as the isolated remnants on Dartmoor (e.g. Wistman's Wood, Fig. 6.5);
3   that most of the woods existing today are of recent origin, having been planted as screens or cover for game.

This last assumption probably represents an over-simplification. Thus the ash woods on the broken rock surfaces of limestone gorges

Figure 6.5 ANCIENT CLIMAX SITUATIONS IN BRITAIN. Oak wood on Dartmoor, screened on the west, north and east by high moorland, where the trees emerge from the shelter of surface rocks which protected them during their early growth. Note the dense covering of epiphytes on the branches. Man is unlikely ever to have planted a wood in such an environment and this is probably a relict of the ancient Dartmoor Forest. Wistman's Wood, Two Bridges.

(e.g. Cheddar and Clifton) are stable climaxes which are likely to have had a continuous existence from before the time of the Neolithic agricultural developments, the terrain having been unsuitable for grazing or the propagation of crops since its formation. Similar pockets remaining from extensive ecosystems could well be more widespread than is commonly believed and the subject requires further study.

The trend is for the climatic climaxes that have been set up by planting, and those of a corresponding nature which have developed by succession, to differ less in botanical than in zoological features. However, their response to any change taking place, after comparative stabilization has been reached through environmental influence, is generally so slow as to limit their value in the investigation of ecological principles. Deflection experiments, wherein organisms are removed to other situations to test their apparent dependence on factors operative in the climax, can provide useful information (see also Chapter 1) otherwise much of the work is apt to be merely descriptive. Subclimax situations, however, have looser stability and are valuable on at least two counts: they are more accessible than climatic climaxes, and they react so quickly to changes in external conditions as to provide material for experimentation over relatively short periods.

6.4 The grassland subclimax

The British climate is particularly suitable for grassland which, in its various forms, is the most widespread of our subclimax ecosystems. Its stability, sometimes maintained by fire, is more commonly maintained by biotic factors of human derivation, such as mowing and grazing, whereby growths from the seedlings of woody plants are kept down. Those herbaceous colonizers which take over survive as a result of various adaptations.

Two familiar examples are, the growing points below the soil, found in many sward grasses, and a rosette form of leaf mosaic adpressed against the ground with the growing point so close to the surface as to miss the destructive effect of the cutters, found in plantains and such dandelion-type composites as hawkweeds. Less familiar are the poisonous and distasteful sap avoided by grazing animals (e.g. *Ranunculus bulbosus*), regeneration from fragments of axillary runners (e.g. *R. repens*) and regeneration from cut pieces of foliage and stem which, in certain conditions, give rise to adventitious roots and ultimately to mature assimilatory growths.

The common daisy, *Bellis perennis*, is an example of the last (see page 109): further investigation might reveal others. Survival, and the degree of disturbance, are obviously connected. The trampling by animals, and the depth into the soil to which cutting is effective, are two features in this connection.

It is noteworthy that dandelions can eventually be eliminated from a sward by sheep, which bite into the ground so as to cut off the growing regions near the crowns of tap roots of primary plants as well as those of any secondary developments which arise from the traumatic break-up of the roots themselves.

Some botanical features of grassland ecology

Investigations of effects of human manipulation on lowland grass communities have been made by Sinker[69], who set up a mowing experiment in spring 1964 at Preston Montford Field Centre, Shropshire, where a protective fence against the larger herbivores had been erected around a field under permanent pasture marked out into nine plots, each 8×3 m, arranged in a Latin square:–

C   B   A
B   A   C
A   C   B

and where each type of treatment was given once in each row and once in each column. The As were cut fortnightly with a mower and the clippings left on the ground; the Bs were scythed once a year at midsummer and the cuttings removed; the Cs were left uncut as controls.

Four years after the start of the experiment a combined NPK fertilizer in granular form was applied to one end of each plot under cutting treatment. Assessed by counts of the number of leaves per dm^2, effects on white clover, *Trifolium repens*, of the different treatments showed up rapidly. Counts were also made of the changing densities of six species of grassland weeds – bulbous buttercup (*R. bulbosus*), ribwort plantain (*Plantago lanceolata*), sorrel (*Rumex acetosa*), daisy (*B. perennis*), dandelion (*Taraxacum officinale*) and common cat's ear (*Hypochoëris radicata*) – measured in terms of the number of plants (=rosettes) per unit area. Here a standard sampling technique was used in which wire-framed quadrats (dm^2 and ft^2) ($0.093 \, m^2$) were placed according to pairs of random

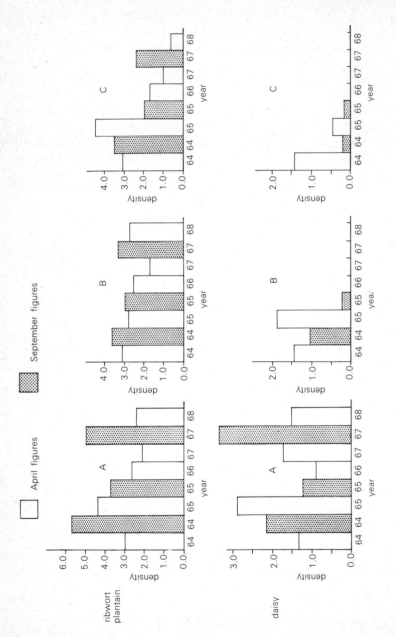

numbers giving a grid reference from axes marked by measuring-tapes along the sides of each plot. Ten 1 ft² (0.093 m²) quadrats per plot gave a sample size of 4 per cent by area, and twenty 1 dm² quadrats a sample of 1 per cent. Histograms in Fig. 6.6 summarize the variations in densities of two of the weeds, plantain and daisy, under cutting and control regimens which Sinker's students recorded during the five-year period 1964–68.

The striking difference in response culminated in the total disappearance of daisies from plots under B and C treatment by the spring of the third year. Results from the A and B plots broadly conformed to the findings of students working with the present writer, who were mainly concerned with the relationship between the activity of horses and the dispersal of daisies, but who considered also the differential effects of contrasting mowing regimens in lawns, where the clippings were either left on the surface or removed (see page 109). An additional factor put forward by Sinker was the difference in morphological plasticity of the two species in response to environmental changes–plantain gave rise to leaves from 5 cm to 46 cm long and from a horizontal to vertical aspect, but daisy showed so little capacity for varying the size or aspect of its foliage as to be at a disadvantage among the rank growths in the control plots. In contrast, sorrel and bulbous buttercup survived partly because of their ability to elevate leafy stems above the surrounding vegetation. Sinker has emphasised that the cheapest way to destroy daisies in a lawn is to refrain from mowing it for two years.

Field studies of this kind reveal relationships between the morphological features of a species and its ecological response.

Counting rosettes, an imprecise method of assessing plants quantitatively, is applicable only to growth forms which can clearly be distinguished as individual units, but it has the advantage of being relatively speedy and of avoiding the need for rather tedious statistical checking, and it can serve for all but the most sophisticated work. Where precision is called for, either harvesting and weighing or measurements of cover value by means of a point quadrat frame, have greater validity (Ashby[3]).

Fig. 6.6 ECOLOGY OF GRASSLAND MANAGEMENT. Changes in the densities per ft² of two rosette-forming weeds. The letters refer to the Latin square in the text. Adapted from Sinker[69]. Preston Montford, Shrewsbury.

Grasses, the main components of the subclimax, were found by Sinker to show differentiation between the various treatments and also between these on the one part and the grassland outside the enclosure on the other. The outside pasture was subjected to grazing, manuring and the application of fertilizer: here perennial ryegrass, *Lolium perenne*, was dominant and the same grass survived in small amounts in A but virtually disappeared from B and C; Yorkshire fog, *Holcus lanatus*, became dominant in A and co-dominant in B; and cocksfoot, *Dactylis glomerata*, became the other co-dominant in B and the dominant grass in C.

Such changes, noticeable even on casual inspection, can lead to programmed treatment calculated to bring about the dominance of a particular species from any given starting point.

Some zoological features of grassland ecology

Populations of certain invertebrates selected for quantitative determination on a school playing field in Worcestershire indicated a close connection between the animals on the one part and, on the other, mainly the cutting regimen to which various parts of the subclimax were being subjected.

Environment. The soil was slightly alkaline. The field, along one of its flanks ended in a shallow dry ditch beyond which rose a bank about 76 m in length planted along the top with bushes of holly, *Ilex aquifolium*. Until 1966, the herbaceous material in the ditch, and on the slope of the bank had been scythed several times a year: after the mid-1960s, stinging nettle, *Urtica dioica*, was removed periodically from the ditch by digging, the herbs on the bank were cut in May once a year, but the main area of the playing field continued to be close-mown throughout the growing season. The chief components of the playing-field sward were perennial ryegrass, *Lolium perenne*, and smaller cat's tail, *Phleum nodosum*, as co-dominants. Where the mowing machine oscillated laterally, a narrow strip developed on level ground between the close-cut sward on one side and the ditch and bank on the other: this became colonized by *Holcus lanatus*. On the bank, *Dactylis glomerata* was eventually the tallest grass with most cover.

An experimental area was included in the games field, where worn turf near a former hockey goal-mouth had been dug out and repaired by planting a few leafy stolons of one of the forms of velvet bent, *Agrostis c. canina*, imported from the United States. To impart an acidic pH to the substrate and inhibit competition from such vigorous invaders as *Lolium perenne* with affinities for basic conditions, the ground here was given a preparatory dressing of two sulphates, ferrous and ammonium, and this was extended as growths from the initial stolons expanded: in four years the species came to dominate about 22 m². Peripheral changes included replacement of the original sward dominants by red fescue, *Festuca rubra*, and smooth meadow-grass, *Poa pratensis*, both of which arrived naturally. After 1971, maintenance of the experimental area was terminated, the grass having been found to be so intolerant of drought as to be eroded by mowing in dry weather, and to be unable adequately to stand up to heavy treading by players.

The mild, open weather of winter 1972–73 promoted the development of particularly extensive and dense patches of mosses in localities where the grassland was partly shaded and stayed damp: there were two principal species, *Pseudoscleropodium* (= *Brachythecium*) *purum* and *Rhytidiadelphus squarrosus* (*Hylocomium squarrosum*).

This gave six situations for investigation: holly hedge, *Dactylis* bank, *Holcus* strip, field sward, mossy turf and *Agrostis* area. The ditch was disturbed so often that stability remained unachieved.

Estimations. Five groups of invertebrates were considered, and these can be regarded as comprising two categories in terms of the methods used to estimate them:
(a) enchytraeid worms, nematodes, tardigrades and mites obtained from material by funnel extraction;
(b) earthworms, counted by occupied burrows, and spiders, either counted as individuals captured or as occupied webs.

Apart from investigation of the moss-infested parts, made only in February-March before the mosses had been removed by raking, most of the work was done periodically during the previous summer and autumn.
(a) Both water funnels (Baermann) and dry funnels (Tullgren) were used (Chapter 8, Fig. 8.1), whereby cryptozoic animals were driven from turf-with-soil by heat and illumination directed from above. A bulb-coring tool, diameter 6.25 cm, was used to cut out corresponding volumes of material flush with the top edge of the corer: a single sample measured 1/2 920 000 of a hectare.
(b) By using insect nets with the same size of aperture and the same size of mesh, swept 10 times in similar arcs by the same

operator with the arm fully extended, captures were obtained under roughly equivalent conditions. The method is particularly suitable for taking small spiders. Webs of the larger species – the tunnel snare of *Tegenaria atrica* and nursery tent of *Pisaura mirabilis* – were sufficiently conspicuous for counting. Numbers of earthworms were taken as the sum of those burrows enclosed by a wire quadrat-frame 45 × 30 cm thrown 10 times at random in a given situation. Only burrows plugged by leaves were counted, and these were assumed to be occupied: worms taken at samples of such burrows all proved to be *Allolobophora longa*, from which the presumption was that they reflected the universal trend at this site.

Results are tabulated as maxima in each situation, irrespective of the time of year when they were obtained. Numerically, spiders were important members of the fauna. Although some of the smaller kinds are common astronauts and had become scattered over a range of situations, the predilection of seven was clear for defined parts of the surroundings. Thus, sheet webs of *Tetrix* were restricted to holly bushes, tents of *Pisaura* to rank herbage on the bank, snares of *Tegenaria* to exposed holly roots, while *Araeoncus* and *Theridion* were commonest in relatively untrimmed foliage of *Holcus* along the margin, and *Erigone dentipalpis* and *E. atra* were really abundant only where the grass had been closely mown. Among spiders observed frequently, but excluded from numerical estimates, the lycosids (wolf spiders) were paramount.

The measurements point to three things: the richness of invertebrate forms of life in one of the commonest of our subclimax ecosystems, the variability of their distribution, and the need for autecological investigations to explain this unevenness. As they stand, the numbers include too wide a range of time and relationship factors for hard-and-fast conclusions: for instance, they do not take account of interspecific competition before the maxima developed.

Population maxima of selected invertebrates in grassland subclimax
x . . . situations not examined

Number of animals, webs or burrows in	Holly hedge	*Dactylis* bank	*Holcus* strip	Field sward	Mossy turf	*Agrostis* area
Arachnidae						
Tetrix denticulata	47	0	0	0	0	0
Pisaura mirabilis	0	11	0	0	0	0

Numbers of animals, webs or burrows in	Holly hedge	*Dactylis* bank	*Holcus* strip	Field sward	Mossy turf	*Agrostis* area
Tegenaria atrica	13	0	0	0	0	0
Erigone dentipalpis } *Erigone atra* }	x	77	74	330	127	34
Theridion bimaculatum	x	82	144	7	20	12
Areoncus humilis	x	88	189	41	12	9
Lumbricidae						
Allolobophora longa	x	2	4	15	4	3

Millions of animals per hectare (from funnel extractions of standard soil cores)

Enchytraeid earthworms	x	37.0	37.0	48.0	48.0	0
Nematodes	x	2.5	37·0	100·0	125.4	58.0
Tardigrades	x	0	2·5	26.0	75.3	4·9
Mites	x	37·0	2·5	23·5	37·0	8·6

Vegetational succession towards woodland climax

Features of succession commonly appear in any grassland subclimax Observations maintained throughout the eight years 1965–73 of a lawn in the Malvern district revealed stages in colonization from pioneers of newly exposed soil to arborescent plants.

A surface drain excavated in the ground as a shallow trench running alongside one edge of the lawn was dug over so frequently as to be colonized only by annuals, many of which were essentially ephemerals–the five commonest species were, shepherd's purse, (*Capsella bursa-pastoris*), red deadnettle (*Lamium purpureum*), sun spurge (*Euphorbia helioscopia*), groundsel (*Senecio vulgaris*) and annual meadow-grass (*Poa annua*). The greater part of the area had remained under grass for at least two centuries and, as far as obtainable records went, had been cut frequently during the growing season. It had formed a particularly fine turf in the main lawn, dominated by two grasses, common bent, *Agrostis tenuis* and red fescue, *Festuca rubra*, both capable of withstanding close mowing, attacks by insects, and adverse climatic conditions. A wide strip along the border farthest from the drainage trench had been subjected to infrequent cutting from the start: here the changes had resulted in the dominants becoming *Lolium perenne* and crested dog's-tail, *Cynosurus cristatus*. Every spring, seedling trees germinated both in the main lawn and the border strip: almost exclusively there were three species – ash, *Fraxinus excelsior*, lime, *Tilia europaea* and sycamore – all arising from fruits adapted for dispersal by wind and all represented in nearby hedgerows by trees at the flowering stage. See Fig. 6.7 for this late phase in succession.

Fig. 6.7 ECOLOGY IN THE SMALL GARDEN. Phase covering eight months in succession of grassland towards forest. 1 Ripe sycamore fruits on tree, 2 *Autumn*, sycamore samara on mown lawn with the seed inserted between grass leaves by the fruit's spinning fall. 3 *Early spring*, germinating sycamore elevating and discarding the winter-rotted husk of its samara; 4 eight seedlings established in 300 cm² of unmown lawn. 5 *Late spring*, most advanced stage reached by young sycamore before destruction by mowing. Length of samara, 4.5 cm. Malvern, Worcs.

From 1967 onwards, the strip was divided into two parts, approximately equal—one half was scythed in spring sufficiently for the seedling trees to be destroyed; the other half was left uncut. By 1969, *Cynosurus cristatus* had declined so much that it had given place to *Dactylis glomerata* as the co-dominant with *L. perenne* in both halves. By 1971, *Dactylis* had become the dominant in the scythed half whereas, in the uncut part, it was sharing the ground with *Holcus lanatus*, and the ratio of the second greatly increased during 1972. By this time, sycamores were the only trees surviving in the area no longer being scythed. They had become small saplings, had passed through their primary phase of epigeal cotyledons and secondary stage of ovate foliage, and had acquired a dense canopy from tertiary palmate leaves. Probably as a result of shading, they had obliterated most of the later seedlings of their own species.

Three of the noteworthy features in this situation are:

(a) Sycamore casts more shade and restricts irrigation of under-lying ground more than any of the other trees developing in the area;
(b) Sycamore, an introduced species, is an increasingly successful invader of what, at one time, were regarded as stable forms of climax communities dominated by indigenous deciduous trees;
(c) *Holcus lanatus*, a particularly adaptable grass, is common as a member of the shaded herbaceous layer in many deciduous woodlands, and its subsequent history in the Malvern situation would be interesting to follow, although the present book went to press before this could be achieved. Further work along these lines is self-evident for projects at sixth form level and above.

Some effects of grazing animals

During the six years 1966–71, students in the Malvern district investigated the effects on grassland of grazing stock—horses and sheep—turned down from time to time in the same two fields for feeding and, in the case of horses, for jumping practice. The manner of grazing differs: horses merely cut off the emergent vegetation whereas sheep bite so close to the surface as frequently to swallow soil. (Cattle, not considered here, roll the tongue around vegetation and cause much disturbance by tearing it and pulling it up in clumps.) Observed effects included destruction by feeding, disturbance by movement, and results of excretion: investigations were confined to the first two.

105

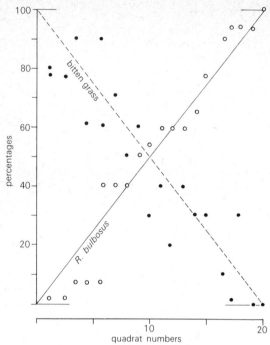

Fig. 6.8 ECOLOGY OF THE GRAZING ANIMAL. Percentages of grass leaves with tips bitten by horses in 20 quadrats, each 30 × 40 cm. Results plotted according to amount of cover of *R. bulbosus*, the most poisonous of the grassland buttercups. Note inverse relationship. Malvern, Worcs.

Feeding. Wired quadrat frames, 30 × 40 cm, thrown at random, revealed *Ranunculus bulbosus* to be the commonest buttercup. When horses alone were the grazers, measurements taken from the beginning of March onwards showed the percentage of grass leaves with bitten tips to be inversely related to the amount of *R. bulbosus* cover per quadrat (Fig. 6.8). Earlier, in mid-February, when new growths began to appear in the grassland components and the first mowing of the year was carried out on nearby lawns, the relationship was less clearly defined, and even some of the *R. bulbosus* foliage was found to have been bitten. Grazing behaviour, closely watched, indicated that the horses showed less discrimination at this time than later in the year when they avoided *R. bulbosus* altogether and, in so doing, missed the nearby grasses. Whether this is due to later increase in the distasteful and toxic properties of the buttercup, to its increasing effect chemically on adjacent grasses, or to a horse's

Fig. 6.9 ECOLOGY OF THE GRAZING ANIMAL. Pasture grazed by horses which have left a large clump of thorn-apple (*Datura stramonium*), area 120 cm^2, shown in foreground. Unlike the three grassland buttercups, *Datura* is an annual. All its parts are narcotic and very poisonous. Its seeming scarcity is one expression of its erratic appearance: evidently the buried seeds retain their viability for many years and germinate if brought to the surface by disturbance. Horses may have had two functions here, their scuffing feet disturbing the soil and their selective feeding cutting the adjacent grass and thereby promoting the weed's expansion. Malvern College grounds, Worcs.

behavioural urge to take succulent young shoots in general irrespective of species, remain open questions.

A noteworthy feature is the rapidity with which *R. bulbosus*, the earliest of the grassland buttercups to flower, dies back after going to seed. In autumn the plant produces fresh growths which overwinter; and this is the foliage which horses graze to some extent at the very beginning of the new growing season. In one year (1968) the 'hooved locusts', sheep, replaced horses in the fields when *R. repens*, the least toxic of the buttercups, was found to be torn by grazing in every month and the sheep were seen to take practically any sort of green vegetation, including *R. bulbosus*.

Slade[70] claims that removal of buttercups from pastures increases their stock-carrying capacity and also results in increased fattening of the beasts.

Disturbance. During 1970–71, both fields were used to train horses for hunting and steeplechasing, when jumps were set up in positions where they remained unchanged throughout the two-year period. There were two groups of horses:

(1) children's ponies fitted with smooth shoes for work on flat ground, taken only over low hurdles and confined to the smaller field;

107

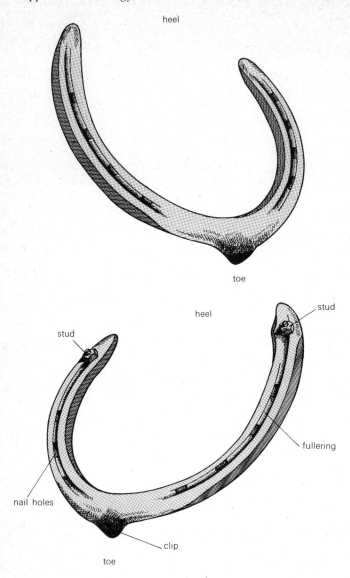

Fig. 6.10 ECOLOGY OF THE JUMPING ANIMAL. The two types of horseshoe used for the project summarised in figure 6.11: *above*, smooth shoe; *below*, shoe fitted with two studs. Scale $\times \frac{2}{3}$.

(2) horses with studded shoes designed to reduce slipping, which vaulted over higher obstacles and which exercised only in the other field. Fig. 6.10 compares the two types of horseshoe.

In the second year, clumping of inflorescences of daisy, *B. perennis*, became noticeable on the far side of the jumps where the smooth-shod ponies alighted and these extended along the tracks taken by the animals running on the flat but diminished with the distance from the landing area: a similar trend showed where the horses with studded shoes exercised but it was less apparent to the eye. By counting the number of daisy rosettes within wire-framed quadrats laid down at intervals, comparisons (Fig. 6.11) were made along three lines diverging from a hurdle in the general direction taken by the jumpers after recovery. Measurements suggest the hypothesis that the ponies' forefeet in smooth shoes slipped forward slightly on landing so that they acted as chisel-shaped cutters and, from existing daisy plants, detached fragments which became established further along the course as vegetative units eventually reaching the flowering stage.

The hypothesis was tested by cutting with a knife pieces from the vertical stock of daisy rosettes, severing leaves at the petiole, and planting out the parts in seed trays containing compost made up from 1 part horticulutral peat, 1 part sterilized loam, 2 parts coarse sand, and traces of superphosphate and powdered chalk. When kept at room temperature the detached leaves, and those fragments of stock bearing axillary buds, were found sometimes to give rise to adventitious roots and to set up new plants which eventually flowered. This accords with what has been observed in lawns, namely that an area cut by a mower without a grass-box tends to develop more daisies than one from which the cuttings are removed. Factors affecting establishment included those which were, (1) seasonal and (2) positional. Thus, in (1), those fragments taken during the period of most vigorous growth (mid-February until the cutting of the first hay crop in the district) rooted most readily: in (2) where leaves were used and roots formed at the cut surfaces of the petioles, those which came to lie with their petioles against the ground were the ones which most commonly survived.

As it stands, the project suggests additional work. This might include investigation of the relative efficacy of fragmentation and fruit dispersal in distributing daisies, and of the influence of tramp-

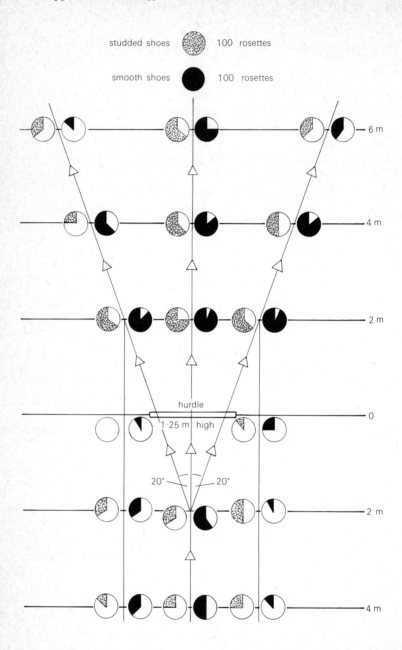

studded shoes   100 rosettes

smooth shoes   100 rosettes

6 m

4 m

2 m

hurdle
1·25 m high

0

20°   20°

2 m

4 m

ling and abrasion on their survival among other sward species. In the field, the possibility cannot be ruled out of hooves breaking up fruiting heads and scattering the seeds, notably in spring when the inflorescences normally attain their maxima; nor can the possibility that daisy rosettes are favoured under certain conditions of mechanical pressure compared with some of the taller grassland components; but whatever the actual sequence of events, it is clear that animals, by movement alone, can effect changes in their habitat.

6.5 *The ecological niche*

An organism's habitat is simply the place where it lives. Its ecological niche is the combination of those particular environmental conditions whereby, in the struggle for existence, it obtains the three basic requirements for biological success: (1), water, together with food of the right sort and in sufficient quantity (or, in autotrophs, the means whereby food can be synthesized); (2), shelter; and (3), a site in which to breed.

The ecological niche implies relationships, including those of competition. It also implies spatial features, although it would be an over-simplification to equate habitat and ecological niche, just as it is inaccurate to visualize the habitat of a male robin simply as a territory delimited in the horizontal plane by hard-and-fast boundaries.

In fact, the writer has personal observations of a bigamous male robin, identifiable in the field by colour rings, which had two mates and two nests simultaneously in one season, the nests beings some 150 m apart, when the male had to pass through part of the territory of another pair of robins on its way from one of its responsibilities to the other, a hazardous journey undertaken many times each day and involving, among other things, the defence of two territories. Consideration of the habitat of this individual and of its ecological niche raises problems of special interest.

The case already mentioned of the rotifer *Callidina symbiotica*, which is practically confined to the small reservoirs at the leaf bases

---

Fig. 6.11 ECOLOGY OF THE JUMPING ANIMAL. Establishment of daisies evidently promoted by ponies vaulting a low hurdle set up in grassland. Numbers of daisy rosettes in quadrats 30 × 40 cm compared for two groups of horses, fitted respectively with smooth and studded shoes, exercising separately in similar fields. Malvern, Worcs.

of the liverwort *Frullania dilatata*, has been presented as an instance of a microhabitat (page 57): the restricted size and highly specialized nature of such an environment point to an animal having a peculiar ecological niche. Examples cited below are those of organisms which, in their activities, encompass a larger amount of space.

Ecological niches in a closed plant community

These are exemplified by herbs on the floor of a climax wood on light acid soil and dominated by durmast oak, *Quercus petraea*: the

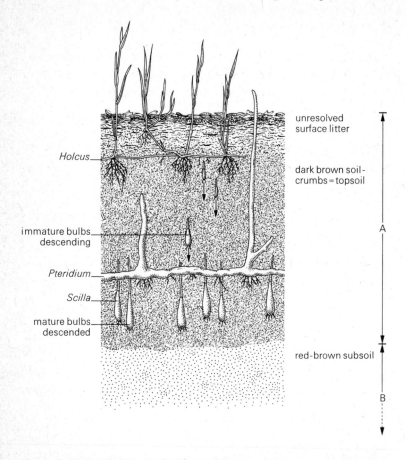

Fig. 6.12 ECOLOGICAL NICHES IN THREE WOODLAND HERBS. The subterranean organs of *Holcus*, *Pteridium* and *Scilla* become so distributed as to exploit different levels of the 'A' horizon of brown forest soil.

three commonest herbs in many woods of this kind are Yorkshire fog, *Holcus lanatus*, bracken, *Pteridium aquilinum* and wild hyacinth, *Scilla nonscripta*. In places their aerial shoots are so close together as to suggest interspecific competition below the surface for water and soil nutrients, but a profile cut through the woodland floor shows the root systems to be stratified at different levels so that different resources are utilized: all lie in the *A* horizon; the bent stems of *Holcus* stretch along the interface between the humus-rich topsoil and the unresolved surface litter and its culms project through this debris into the light; *Pteridium* is deeper and its rhizomes generally run through the upper levels of the topsoil; and the bulbs of *Scilla* are grouped near the base of the topsoil (Fig. 6.12). Although further investigation is needed on the mechanisms whereby these levels are attained and the positions held, observations indicate that, in *Scilla*, the small bulbs arising from its seedlings lie first in the litter layer and then, before flowering begins, develop contractile roots in their second and third year which pull them down to the level they come to occupy permanently.

Ecological niches in pisciculture

The eastern Mediterranean being poor in seafish, Israel is developing pisciculture on a large scale (Coleman[13]). Here food fish are reared in ponds constructed in the coastal strip on the site of drained swamps where the salinity of the water supply renders the land unfit for most kinds of agricultural exploitation. Species commonly stocked in the rearing ponds include the carp, *Cyprinus carpio*, silver carp, *Hypophthalmichthys molitrix*, grey mullet, *Mugil cephalus* and various members of the genus *Tilapia*. Except for the grey mullet, which is really a marine fish whose fry enter estuaries, all can be regarded as typically freshwater forms which are capable of withstanding brackish conditions.

Such public health problems as poisoning do not arise since pond-cultured fish are free from mercury and other toxic substances found in many seafish. A watch has to be maintained to see that disused ponds do not harbour the snails which are hosts for the parasite causing bilharzia in man, and that they do not degenerate into the kind of weed-infested situations to which the malarial mosquito might return and breed. To this end, the coypu, *Myocastor coypus*, has been successfully introduced to keep weeds in check.

Prepared foods are widely used and yields have been increased by feeding various by-products from other branches of the Israeli economy, including kitchen waste and poor grain, and some fish farmers give their carp chicken heads and legs. A striking feature has been the success following the stocking of small ponds by large populations of mixed species. Such a procedure might be expected to cause significant losses through predation but, in practice, it has resulted in higher returns of consumable protein brought about by the fact that the different species come to occupy different ecological niches, with an accompanying diminution in interspecific competition.

Organisms in the foregoing cases are restricted in their movements; those in the section which follows are flying insects which roam more freely.

Ecological niche in digger wasp
Two species of the genus *Mellinus* are among the digger wasps occurring in Britain, the field digger, *M. arvensis*, with a general distribution, and the sand digger, *M. sabulosus*, confined mainly to maritime situations where it may replace *arvensis* locally as the commoner *Mellinus* wasp. Their habits are similar—each excavates a burrow in soil and from this, at various intervals, up to 10 cells lead off stocked with four to nine flies paralised by stinging, which serve as food for the offspring deposited as an egg on the topmost prey.

Competition might be expected between these closely related species having comparable needs, but the writer has found *M. sabulosus* to take flies which frequent particularly the inflorescences of wild carrot, *Daucus carota* and of certain grasses (e.g. *Aira*, *Agropyron* and *Catapodium* spp.) which are often plentiful near the coast, whereas *M. arvensis* is more catholic in its selection of prey. Furthermore, for reasons which he has been unable to discover, *M. sabulosus* males utilize the foliage of coltsfoot, *Tussilago farfara*, as communal resting places. Commonly, wild carrot abounds in coastal areas and coltsfoot in the talus slope at the foot of a crumbling sandy cliff. The newly exposed sand on the face of such a cliff (or quarry) is the kind of material and situation in which *M. sabulosus* commonly digs its burrow, as against *M. arvensis* which seems to utilize a more varied range of situations and light soils. Those features of its ecological niche noted here (Fig. 6.13) are also found where its range extends inland.

114

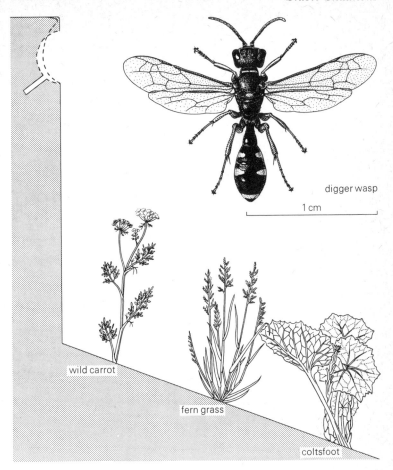

digger wasp

1 cm

wild carrot

fern grass

coltsfoot

Fig. 6.13 ECOLOGICAL NICHE IN DIGGER WASP. Environmental features utilized by the wasp *Mellinus sabulosus*. Walton-on-the-Naze.

Quantitative comparisons between ecological niches of organisms obtaining their necessities in closely similar ways have been made between parasitic insects inducing galls on oak leaves, initially by Askew (see Lewis and Taylor[42]) who used material from Wytham Woods, Berkshire, and subsequently by the present writer with material from Hoggat's Wood near Bishop's Stortford, hedgerows near Malvern sewage farm, and Tewkesbury.

One Approach to Ecology

Ecological niches in gall wasps

In Britain, four gall wasps (cynipids) of the genus *Neuroterus* cause the familiar flat, disc-shaped galls (spangles) on the underside of leaves of pedunculate oak, *Quercus robur*, and durmast oak, *Q. petraea*, particularly on the former, the distribution of which tends to be more southerly and therefore in milder regions more favourable to insects generally. The four are, common spangle (*N. quercus-baccarum = N. lenticularis*), cupped spangle (*N. tricolor = N. fumipennis*), silk-button spangle *N. numismalis = N. vesticator*) and smooth spangle (*N. albipes = N. laeviusculus*). Fig. 6.14 shows their external appearances. As in other oak cynipids, a spangle is one of two galls in the life cycle of the same species: it is an over-wintering stage caused by a female which originates from sexual fusion and acquires a certain amount of protection inside the spangle through the latter becoming detached from the foliage before the leaves themselves fall from the tree. Eventually these cover the spangles already lying on the ground so that they are concealed from predators (game birds, etc.), blanketed from frost and kept damp.

In spring, a single female emerges from each spangle. These are agamic females which lay parthenogenetic ova – as a rule in the buds of oak, although *N. tricolor* commonly inserts its eggs in the expanding leaf-blades: here different galls develop rapidly in summer, and from them both sexual females and males arise which mate and produce the fertilized eggs deposited into the smaller veins through the lower epidermis of the leaf and which hatch into the larvae of the agamic females causing the spangles to form. The life cycle of a *Neuroterus* is given in Fig. 6.15: it exemplifies the ecological principle that, where two methods of reproduction occur in the same organism, there is a tendency for genetical interchange to follow a favourable period and precede one of adversity, whereas asexual multiplication is associated with a time of relative abundance.

Young trees near the periphery of a woodland and scrub oaks in a hedgerow are particularly susceptible to attack, a single host and even one leaf often bearing spangles of several species; and estimates of colonization of a single leaf is that it may occasionally result in coverage of up to 80 per cent of the under-surface (Askew in Lewis and Taylor[42]) and exceptionally up to 90 per cent (Darlington). Such densities imply competition for *lebensraum*: the competition may be both intraspecific and interspecific.

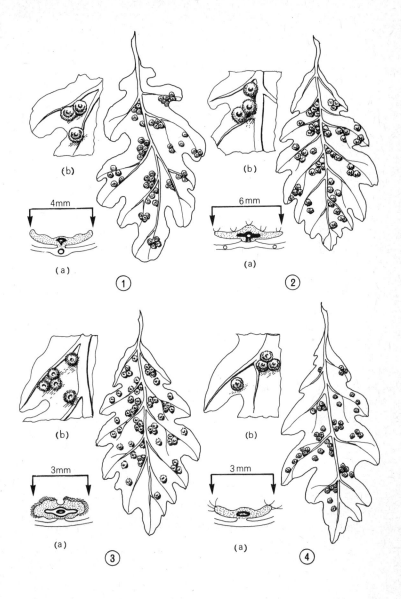

Fig. 6.14 ECOLOGICAL NICHES IN GALL WASPS. The agamic galls (spangles) of four species of *Neuroterus* on oak, (a) section, (b) surface view: 1, *N. albipes*; 2, *N. quercus-baccarum*; 3, *N. numismalis*; 4, *N. tricolor*.

117

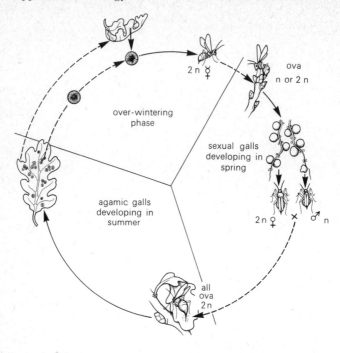

Fig. 6.15 ECOLOGICAL NICHES IN GALL WASPS. Alternation of generations in yearly cycle of a typical oak cynipid, *Neuroterus quercus-baccarum*. Galls of the sexual generation commonly develop in the male catkins and resemble red currants. Galls of the agamic generation drop from the tree before the leaves fall and eventually become blanketed by litter, in the humid atmosphere of which they continue to expand by absorption of water.

One feature of intraspecific competition shows up shortly before the spangles drop, when a small dry area forms in the leaf around each of the older galls. Water is important to the development of the spangles and any die which are so young as to be out of phase with the more mature galls but which have become involved by their positioning with the dry patches near the attachments of the latter. (It is noteworthy that fallen spangles continue to increase in size by absorption of water after separation from their host, and that the covering of the leaf debris provides the humid conditions in which expansion by this means can take place.)

The earlier a female of the bisexual generation can lay the better. Records obtained during the five years 1966–70 by the writer's

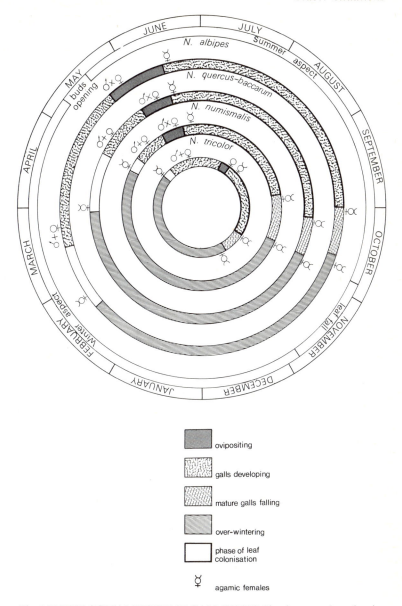

Fig. 6.16 ECOLOGICAL NICHES IN GALL WASPS. Simultaneous chart showing extent to which yearly phases of four *Neuroterus* species overlap. It reflects the significance of differential colonization of the leaf area by the agamic galls. Bisexual galls are initiated in blank sectors of circles.

119

students from a variety of situations in the Malvern–Tewkesbury district are summarised (Fig. 6.16): despite considerable differences from year to year and from one situation to another, the general trends persisted. At one extreme of the yearly period was *N. albipes*, whose galls containing the bisexual generation had developed from infested buds by the third week of March and whose sexual females, issuing from them, began ovipositing as soon as the leaves had expanded: at the opposite was *N. tricolor*, whose bisexual galls only appeared while the leaves were undergoing expansion and whose spangles, initiated by the fertilized products from these, first showed on the foliage at least a month after the spangles of *N. numismalis* and six weeks after those of *N. albipes* and *N. quercus-baccarum*. Such observations point to interspecific competition for the same host and to a diversity of ecological niches and suggest the hypothesis that competition pressures are reduced by one or more of three means: (1) an individual host (tree) is utilized exclusively by one particular species; (2) different areas of a small part (leaf) are utilized by different species; and (3) different regions of an entire host (tree) become colonized by different species. Simple observa-

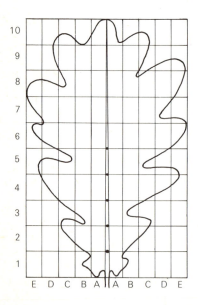

Fig. 6.17 ECOLOGICAL NICHES IN GALL WASPS. Method of subdividing oak leaf for measuring the distribution of spangle galls charted in Fig. 6.18.

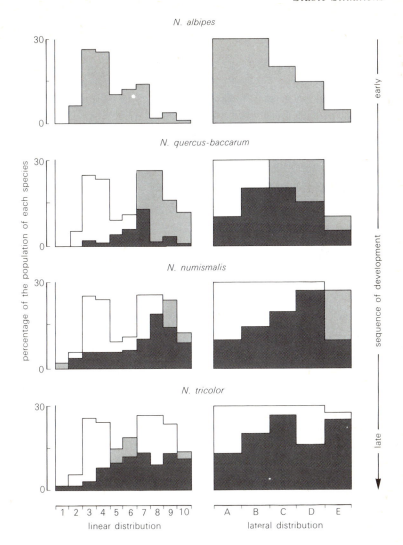

Fig. 6.18 ECOLOGICAL NICHES IN GALL WASPS. Spatial distribution over leaf surface of agamic galls of *Neuroterus*. Histograms actually refer to leaves shared by two or more species, but no difference was found between these and cases where colonization was by a single species.

Note trend for later colonizers to occupy areas away from base and midrib and towards lateral margins and apex. *N. tricolor* has an ill-defined pattern of distribution which evidently reflects a tendency to fill in gaps left by earlier arrivals. Malvern–Tewkesbury.

121

tions show the first to be untenable, certainly in the west Midlands.

The second—differential distribution on a single leaf—has been checked by Askew and repeatedly by our students from foliage collected at random from oaks of various ages in a range of situations within the Malvern-Tewkesbury region. Askew's method of dividing a leaf transversely into tenths, with the addition by the writer's workers of lateral subdivisions into fifths from each side of the midrib, was followed throughout (Fig. 6.17). Results presented as histograms in Fig. 6.18, indicate that, irrespective of pre-colonization, spangles of *N. numismalis* occur almost to the margins of a leaf,

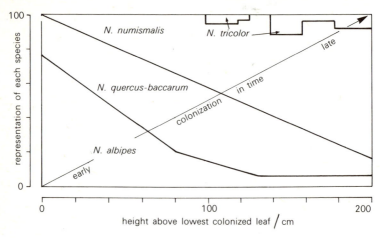

Fig. 6.19 ECOLOGICAL NICHES IN GALL WASPS. Spatial distribution, relative to height in oak tree, of agamic galls of *Neuroterus* spp. Results summarized from histograms. Note tendency for the later colonizers to attain their maxima at the higher levels. Malvern–Tewkesbury.

with *N. albipes* (the earliest arrival) predominating nearer to the midrib, and *N. quercus-baccarum* occurring most plentifully towards the apex. The pattern of colonization by *N. tricolor* (the latest of the four to induce spangles in a season) is the least clearly defined: although the tendency is for the middle regions to be utilized, between the midrib and the lateral margins and between the base and apex, this is a more general distribution than that of any other and suggests the importance of initial colonization by competitors in determining what happens to this species, which seems merely to fill in the gaps they leave. Of interest in this connection is the fact that *N. tricolor* appears to be the rarest of the four within those particular regions studied by the writer. The mid-season arrivals, *N. numismalis* and *N. quercus-baccarum*, more or less coincide in their times of ovipositing; although *N. numismalis*, a little later than *N. quercus-baccarum*, tends slightly towards a diffuse form of distribution after the manner of *N. tricolor*. All three species, *albipes*, *numismalis* and *quercus-baccarum*, retain their pattern of spatial selectivity relative to the leaf's morphological features irrespective of whether or not they come to share the same leaf: evidently such a form of behaviour contributes to success under the pressure of interspecific competition.

The third–spatial distribution over the entire host–has been investigated by collecting, from each of a number of young oaks in a variety of hedgerow conditions, 100 leaves approximately 50 cm from the trunk at successive intervals 10 cm apart, from ground level up to 200 cm. The species were counted at the different levels and calculated as the percentage representation at each height (Fig. 6.19): the original histograms indicated the predominance of *N. numismalis* at the top and periphery, *N. quercus-baccarum* in the middle and *N. albipes* towards the bottom. Again *N. tricolor* behaves as an 'opportunist', filling in gaps indiscriminately over much of the tree. In view of the large size attainable by the host, this is a difficult test to apply more generally, but such extension is needed.

6.6 Autecology and synecology

Although this chapter deals partly with the concept of climax communities, its emphasis is on autecology–the investigation of relationships of an individual species to its surroundings, contrasted with synecology where a number of species sharing the same general environment are considered together. Essentially an autecological investigation is based on the isolation of a single problem and the work towards its solution. Three points inherent in ecological problems have directed the modern slant towards autecology, which increasingly is accepted as the realistic approach where the hypotheses derived from field observations require testing.

1 In a particular situation, its autecological phenomena make up its synecology.
2 Its synecological features have little in common other than those aspects expressible only in descriptive terms.

3   Synecological studies, as such, have proved difficult to examine and assess.

Much of the difficulty arises from the complexity of the subject: the varied situations which comprise an ecosystem and the diversity of organisms involved therein. The value of the work is enhanced where the investigator has set out to make himself thoroughly familiar with one small area – a restricted piece of seashore or a limited part of woodland – and has concentrated his attention on what he knows.

7 Populations and changes in their surroundings

A consideration of the niches outlined in the previous chapter leads, in its turn, to a consideration of some of the factors which influence the distribution and survival or organisms in a given situation.

7.1 Interaction between colonizers

Although sometimes difficult to locate in the field, colonization from zero can be induced experimentally. It has been studied under aquatic conditions by suspending sterilized glass slides in fresh water – in a stream by Gause[26], and in still surroundings by Darlington who used a laboratory tank measuring $120 \times 60 \times 60$ cm containing unfiltered pond water. Its glass sides were cleaned daily of deposited solids. In both investigations, the suspended plates were examined every other day for colonization by algae (Fig. 7.1).

Gause found that the number of algal species increased rapidly and, at their peak, settled down to a fairly constant numerical level: those appearing first were species occurring in the immediate surroundings but, when the plates became fully colonized after a little over a week, the pioneers disappeared and were replaced by others apparently better adapted to the conditions on the slides. Darlington also found a rapid increase initially in the number of algal species leading through succession to stability; but here stability followed a decrease from the earlier optimum. Propagules from a smaller species range might be expected to reach the plates in the second investigation and the mechanism of competition to be affected in consequence.

An obvious relationship between flora and fauna is where feeding is concerned: Darlington studied some of the effects on the algal species by adding large-sized feeders in the form of tropical fishes after vegetational stability had arrived. The tank was partitioned into four sections, one was retained as a control, while into the others were introduced respectively the toothed carps *Xiphophorus helleri*

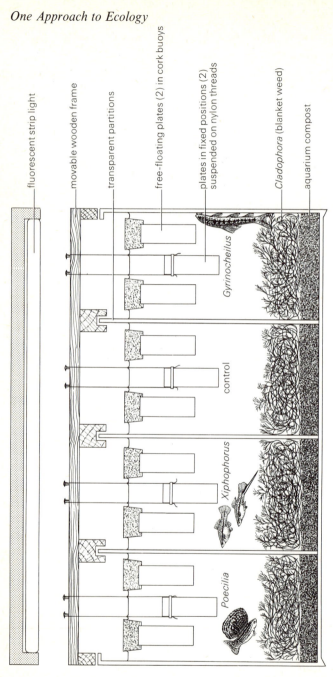

Fig. 7.1 NUMERICAL SUCCESSION ON SUBMERGED PLATES. Arrangement for investigating experimentally some of the conditions affecting algal colonization in still water. Dimensions of tank, 120 × 60 × 60 cm.

(swordtail) and *Poecilia latipinna* (sailfin mollie), and the carp-like *Gyrinocheilus aymonieri* (Thai stone-sucker.) (Fig. 7.2). All take some plant matter. *Xiphophorus* is a predator tending towards the omnivorous habit, *Poecilia* is an omnivore which loses condition unless it receives green matter in the diet, and *Gyrinocheilus* is a herbivore with a suctorial mouth specialized for rasping algal growths from firm substrates. All had to depend for survival on material found in the tank: individuals showing signs of distress were removed and maintained elsewhere. Those which eventually reached a state of balance comprised two swordtails, two mollies and a single *Gyrinocheilus*.

In terms of numerical changes in the species of algae, results from the Gause and Darlington experiments are compared graphically in Fig. 7.3. Differences between the effects produced by the several species of fishes apparently reflect both dietetic differences and differences of accessibility. The small, actively swimming *Xiphophorus* and *Poecilia* tend to remain permanently in midwater during the day and to slip easily in and out of the gaps between the slides: the larger *Gyrinocheilus* is a more sluggish animal which spends much of the time at rest attached to surfaces but which becomes active spasmodically. It seemed to find more difficulty in moving freely between the close-set plates. The slides also supported a microfauna, found by Darlington to consist mainly of ciliates and rotifers, but these proved so sporadic in appearance and transient in survival as to defy correlation between flora and fauna. Appearance of the fishes was followed by the virtual disappearance of these tiny animals, but whether this was the consequence of direct predation, an incidental effect of algal browsing, or some other factor, remains unknown. By the time the fish compartments had become stabilized, the slides in the control section had acquired so thick an incrustation of algae that counting had become difficult.

Even in the restricted volume of an aquarium, the slides were clearly part of a wider environment with influential factors operating outside them. In such communities, actual numbers of individual species or of their populations have little significance in terms of hard-and-fast relationships until the factors determining their establishment have been isolated and assessed separately. They indicate the need for an autecological approach to investigations. Exotic species might be considered to introduce an element of artificiality and, in a sense, they do; but an aquarium, being basically

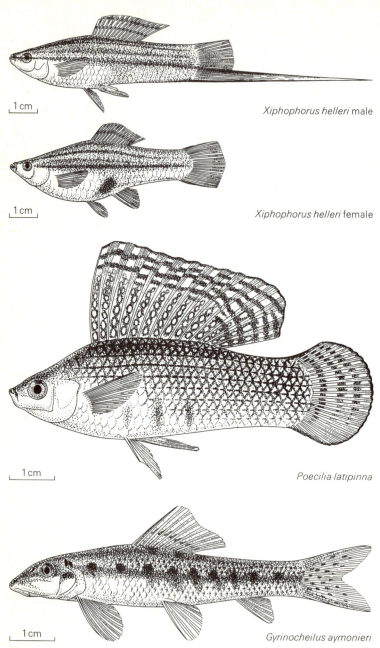

1 cm

Xiphophorus helleri male

1 cm

Xiphophorus helleri female

1 cm

Poecilia latipinna

1 cm

Gyrinocheilus aymonieri

a water garden, is no more artificial than a terrestrial garden–both are valid situations for testing out ecological relationships. The so-called 'toy fishes', which are easily managed, tend to be used insufficiently for such work.

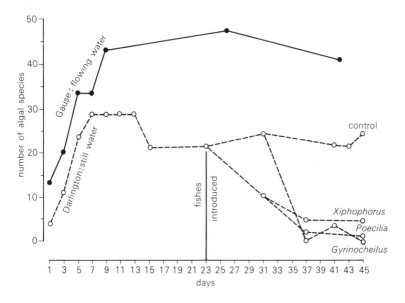

Fig. 7.3 NUMERICAL SUCCESSION ON SUBMERGED PLATES. Changes in numbers of algal colonizers found respectively by Gause and Darlington. 'Seeded' communities set up by the introduction of browsers facilitate the isolation of differentials such as those of diet and accessibility.

7.2 Interrupted succession

So many variables operate in aquatic conditions that the communities developing therein are as diverse as the sites themselves. In the sense that succession implies changes in species which develop until a balance is reached, succession in water is particularly liable to interruption on several counts. Three of these are, the possibility of catastrophic termination through drying up, the restricted volume of water in static conditions, and the influence under conditions of

---

Fig. 7.2 NUMERICAL SUCCESSION ON SUBMERGED PLATES. Three species of warm-water fishes introduced as algal browsers in Darlington's investigation.

flow of events occurring upstream.

Features arising from the the first two are considered.

Adaptations for interrupted succession in aquatic situations, shown by animals distant from the initiation of food chains, are exemplified by the following.

The natterjack toad, *Bufo calamita*, is a predator with three features facilitating colonization of pools which are liable to dry up: its rapid locomotion over ground so that a range of breeding sites may be encountered, its relatively protracted spawning period (18 weeks), and its short period of metamorphosis (6–8 weeks). Typically it metamorphoses in temporary pools on coastal dunes, where its osmotic tolerance allows breeding in brackish waters.

Larvae of the culicine *Aëdes*, indiscriminate feeders on suspended organic matter, include species which emerge from eggs hatching in successive groups at fairly lengthy intervals, and in which the risk of all being killed by premature drying of the surroundings is thereby reduced. (*Aëdes* merits comparison with the goosefoot *Chenopodium album*, a terrestrial pioneer of newly turned soil which is also adapted for shifting conditions. Its copious seeds are of several groups: one-third germinate immediately: if the remainder are buried, then brought to the surface and exposed to cold, up to one-half of these subsequently germinate: nitrogenous radicals appear also to operate in breaking dormancy.)

Temporary pools in wheel-ruts which receive the runoff from manure are the classic habitat of another filter-feeder, the fairy shrimp, *Chirocephalus diaphanus*. Hall[28, 29] finds that the number of generations in a year appears to be related to the frequency with which the habitat dries; that so long as the water remains deeper than 20 cm no development takes place, but that it starts slowly as the depth becomes reduced by evaporation; and that partly developed eggs survive in the dry substrate and resume growth and soon hatch after inundation.

Observations such as those on *Chirocephalus* suggest the possibility of some of the forms earlier in the chain being similarly adapted.

At municipal refuse sites, temporary pools in new wheel-ruts and discarded containers, receiving vegetable debris from household garbage and water from precipitation, undergo colonization from zero and have been found to show the general successional trend in the protozoa:–

130

$$Colpidium \; \rightarrow \; \begin{matrix} Paramecium \\ Vorticella \end{matrix} \; \rightarrow \; \begin{matrix} Loxodes \\ Stylonchia \\ Chilodon \end{matrix}$$

notwithstanding other variables which bring about the formation of as many different kinds of community as there are pools. Frequently succession proceeds to the stage of rotifers and arthropods before evaporation causes population crashes. The situations resembled those in blocked roof guttering, where the energy originates from drifted dead leaves and succession sometimes results in aquatic diptera and emergent vegetation before the water disappears (Mackie[48]). At the refuse tips, where re-flooding occurred the stages in the succession were found to appear more rapidly the second time round and, to some extent, to become telescoped, thus leading to the supposition that re-colonization arose, in part, from propagules surviving in the dried substrate.

This was checked experimentally from comparisons between the sediments in two gutters and three temporarily dry pools in nearby wheel-ruts: distilled water was added to two samples from each, respectively unsterilized and sterilized by heat, exposed to air on a roof, and results compared with developments in pools *in situ*. Two methods were used for comparing the populations of small organisms – counts by haemocytometer and the frequency of their occurrence in water drops.

1   The haemocytometer, a slide with a circular depression 100 microns below the surrounding surface wherein were two engraved lines at a known distance apart, received a drop of the sample: a coverslip was pressed down over this until interference patterns with the surroundings appeared. Thus a standardized volume of fluid was obtained from which numbers were counted between the lines.

2   The maximum frequency was assessed by the number of drops, out of a random sample of 10, in which each of the organisms in suspension was found by microscopic examination. Simple presence ($+$) or absence ($-$) were used to record the larger invertebrates and such diffuse forms as filamentous algae.

Although the more precise haemocytometer measurements might have been expected to supply more significant data than the simple and rapid comparisons made by the water-drop method, in practice this was found not to be the case, and the latter have been used to compile the following table which lists 15 of the organisms appearing within three weeks of re-flooding (June 2–23, 1968).

Number of drops out of ten recolonized within three weeks

Colonizers	I Sterilized substrate Pools	Gutters	II Unsterilized substrate Pools	Gutters	III In situ Pools	Gutters
Blue-Green Algae						
Phormidium etc.	10	10	10	10	10	10
Flagellates						
Chlamydomonas	10	10	10	10	10	10
Chilomonas	10	10	10	10	10	10
Euglena	—	—	10	—	10	—
Vesicular coenocyte						
Botrydium	—	—	+	+	+	+
Filamentous Alga						
Vaucheria	—	—	+	—	+	—
Protozoa						
Actinophrys sol	—	—	5	1	5	3
Colpidium	10	10	10	10	10	10
Paramecium	—	—	10	10	6	4
Vorticella	—	—	3	8	10	10
Coleps	—	—	4	4	6	10
Rotifers						
Philudina roseola	—	—	8	6	8	8
Others	2	4	7	7	10	8
Diptera						
Chironomus	—	—	—	—	+	+
Aëdes	—	—	+	—	+	—

It is evident that recolonization by the majority of the smaller organisms considered here is promoted when conditions within substrata remaining from earlier inundations are not those of sterility. But the true extent to which alternate flooding and drying give certain of these forms an advantage over those arriving from

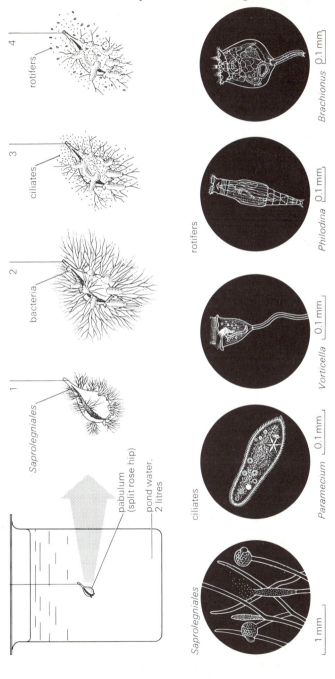

Fig. 7.4 SUCCESSION OF GROUPS IN WATER. Four stages during developments within communities around decaying matter. The pond water had been filtered. 1, Saprolegniales (water-moulds) developing as primary decomposers. 2, Bacteria developing, Saprolegniales reproducing. 3, Ciliates developing, Saprolegniales declining (autointoxication?). 4, Rotifers developing, ciliates declining (predation?).

outside–and for how long the advantage prevails–remain problems for additional project work.

Where the environment is spatially restricted, as in the still water held in a small container, collapse of the dominant organism, and failure thereby of its subordinates, may come about as a result of autointoxication. Frequently this is seen in the communities which start to develop among the hyphae of water-moulds (*Achlya* and *Saprolegnia*) growing as saprophytes on submerged dead matter (split hemp-seed, rose-hip, etc.). The hyphae form a zone radiating from the pabulum. Notable features of the colonization include the establishment of bacteria near the base of the fungus with ciliates (e.g. *Vorticella, Paramecium*) feeding on these, and such predators as *Bursaria* and rotifers devouring *Paramecium*. The situation is that of a microhabitat, where the components are linked by a food-web, the hyphae provide shelter, and the whole occupies a particularly small volume, often less than 1 cm³ (Fig . 7.4).

In such situations, water-moulds are particularly liable to sudden population crashes, their thalli disintegrating within a few hours of visible inception of decline. In general terms, the larger the volume of the surrounding water, the longer the community remains. A physiological characteristic of water-moulds is their sensitivity to low concentrations of solutes in the environmental liquid, probably a result of the low osmotic potential of their cell sap. In small quantities of non-flowing fluid, accumulations of their metabolic wastes presumably soon reach the point at which they become toxic.

7.3 Effects of plant succession on animal communities

Close dependence of animal succession on that of vegetation is a feature of sites where colonization develops on bare ground and the initial plant communities are 'open'. The further the succession proceeds the looser the connection.

In the high Alps, Janetschek[37] has studied a comparatively uncomplicated succession with a steep gradient of physical factors

SUCCESSION OF INVERTEBRATES IN HIGH ALPS, selected from Janetschek. Note that the clearest correlation with plants is between zones I and II, where succession is at an early stage.

Ants
Grasshoppers
Spiders
Later beetles
Quedius alpestris beetle
Nematodes
Earthworms
Lithobius lucifugus centipede
Notiophilus biguttatus
Amara quenseli

Pioneer beetles

Pemphigus similis
Janetschekia lesserti spider
Less specialized collembola
Pardosa wagleri nigra wolf spider
Protisotoma crassicaudata
Isotoma saltans

specialized collembola

Zone O Algae with lichenised algae

Zone I *Poa laxa* with *Cerastium uniflorum*

Zone II *Polytrichum juniperum*
a with *Agrostis rupestris*
b with *Trifolium pallescens*

Zone III *Festuca halleri*
a on glacier soils
b on sunny slopes

Zone IV High alpine meadow association

135

and defined vegetational zones, where bare rock is gradually uncovered by ice and the site becomes progressively colonized by higher plants as soil accumulates. He recognizes four zones of leafy plants. The table on page 135 which gives a simplified summary of his findings suggests that, although correlation exists between these plant zones and the distribution of animals, at best it is only approximate and the clearest demarcation they share is the boundary between zones I and II, i.e. where succession is at an early stage. This incomplete dependence of animal associates on obvious plant zones seems common in certain other ecosystems, notably where arctic or sand-dune animals are under investigation. Evidently the structure and spacing of the soil particles, together with the density as such of the vegetation, can be more important than the specific composition of the flora. Janetschek maintains that, in more advanced alpine conditions, myriapods and some snails are strictly related to vegetation whereas other forms, notably collembolans, are influenced in their distribution by the structure of the soil.

The present study ranged from such specialized pioneers as cold-resistant collembolans feeding on algae to members of the animal communities among plants of the high alpine meadow association. Not all the species were equally restricted in their range: thus the centipede *Lithobius lucifugus* occurred throughout zones II and III, but the spider *Janetschekia lesserti* was so localized as to be essentially an indicator species, that is one pointing the direction of a particular successional trend but otherwise having little significance.

There are opportunities for comparable work under British conditions: the colonization of walls, for example. The desirability of an autecological approach is inherent in the foregoing.

7.4 Effects of plant succession on a single species

One autecological investigation of an animal affected by the changing diversity of conditions apparently brought about by changes in vegetation is that of the ground beetle *Nebria brevicollis* (Fig. 7.5), carried out independently by Paxton, of Haileybury College, and Darlington. This is a cryptozoic form, lurking in crevices under stones, plant litter, etc. and ovipositing here in May–June, a gravid female carrying 20–30 eggs under the elytra and giving rise to offspring which mature in mid-September. It is both a saprophage

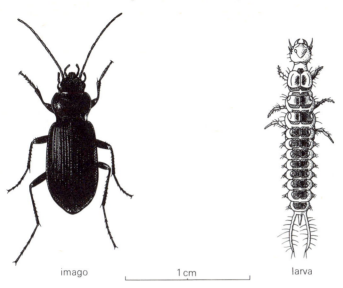

imago   1 cm   larva

Fig. 7.5 AUTECOLOGY OF *NEBRIA BREVICOLLIS*. Imago and larva of beetle.

and a predator, its prey including earthworms and insect larvae. Succession was deduced from stabilized situations – grassland, coniferous woodland and broad-leaved deciduous woodland.

Comparisons between the populations in these situations were derived from captures obtained in unbaited pitfall traps so distributed that each was 300 cm from the one nearest to it. Paxton took his highest numbers in deciduous woodland and found grassland to be intermediate between this and the coniferous plantation, where he caught none. Darlington (summer–autumn 1969) used a total of 80 traps distributed in four groups of 20: his captures are summarized below as percentages of total capture.

Deciduous woodland under oak	57
Mixed woodland under spruce and oak	12
Grassland within 600 cm of deciduous wood	27
Grassland beyond 600 cm of deciduous wood	4

Absolute values, determined by the Lincoln Index (See 8.3), proved difficult at the higher densities because of the small proportion of recaptures here. In mixed woodland and in grassland distant from tree cover, recaptures were nearly 10 of the initial samples and gave populations within the range of the traps of 3 800 and 1 250 respectively: thus, assuming the validity of the listed ratios, the population of all the areas encompassed by the trapping was around the 30 000 mark.

Although colour marking might be used to trace movements of individuals between situations, it would require a complex coding system and was not attempted. The fact that a comparatively high proportion of recaptures was obtained in certain localities–notably where totals were small–suggests a tendency for the insects to remain within restricted areas. Traps were cleared twice daily, morning and evening: that the beetles seemed to be essentially nocturnal in their wanderings was apparent from the differences between the captures at the two times, less than five being taken during the day. Such behaviour reflects a need for concealment as also does the differential results from the breakdown of the grassland into two parts respectively, one near and the other distant from woodland cover.

Paxton determined their peak activity experimentally by recording the interruptions to a beam directed on an infra-red cell by the movements of the beetles. The apparatus and circuit used are shown respectively in Figs. 7.6 and 7.7 and his results are graphed in Fig. 7.8. When there was no visible illumination, he found a marked peak of activity between 2300 and 0600 hours: under white light activity became greatly reduced, although even here the tendency persisted for a peak to develop between the same hours: the use of Wratten colour filters showed the beetles to be particularly tolerant of red, orange and green wavelengths.

That avoidance of light has survival value is self-evident in a species like *Nebria brevicollis* which is taken by diurnal predators including many birds (Collinge[14]). It could also relate to humidity. By hygrometer, Paxton found the relative humidity at 19°C on the ground surface to be 66 in grassland and 67 in coniferous woodland when it was 75 in deciduous woodland–differences which did not correlate with differences in trapping success between the coniferous stand and the grass. Darlington undertook a longer series of measurements, the average relative humidity being 72 in both parts of the grassland when it was 64 in the spruce-oak woodland and 100 in the oakwood stand, thus establishing a closer correlation with the numbers of his captures. Even at night, the humidity under spruce-oak never seemed to reach 100 per cent. When Paxton exposed his

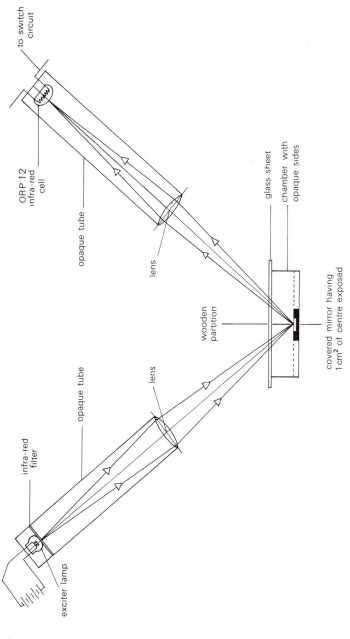

Fig. 7.6 AUTECOLOGY OF *NEBRIA BREVICOLLIS*. Optical part of Paxton's apparatus for recording locomotory movements of insects in the dark by interruptions in an infra-red beam. The same apparatus can be adapted for recording activity under visible illumination.

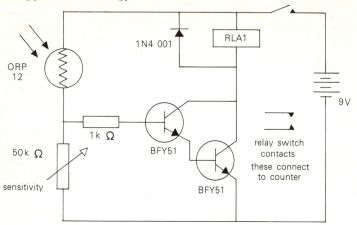

Fig. 7.7 AUTECOLOGY OF *NEBRIA BREVICOLLIS*. Switch circuit receiving impulses from optical apparatus. The relay is incorporated into another circuit so that, each time the contacts are closed by a beetle breaking the beam, the digital clock (numerical counter) advances one unit.

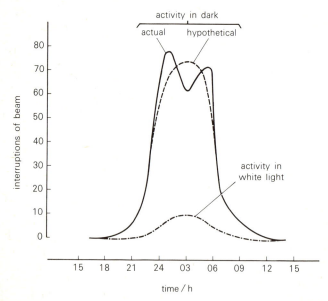

Fig. 7.8 AUTECOLOGY OF *NEBRIA BREVICOLLIS*. Activity in dark and light respectively recorded by the equipment shown in Figs. 7.6 and 7.7. Evidently the beetle has a built-in 'biological clock' which induces a similar rhythm under both conditions, despite the difference in intensity of the activities. The peak develops around 0300 hours.

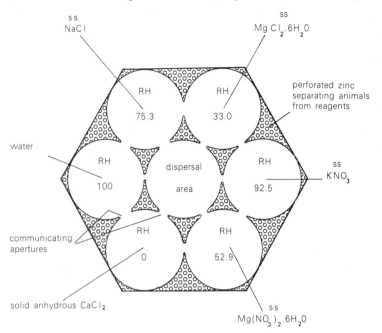

Fig. 7.9 AUTECOLOGY OF *NEBRIA BREVICOLLIS*. Multiple choice chamber, constructed from plastic Petri-dishes, supplying atmospheres of six humidities. Saturated solutions are denoted by ss: humidities are valid for a temperature of 25°C.

material in a dry atmosphere at 22°C, he noted water-loss through evaporation to kill the imagines in 6–7 hours and to render them brittle. By setting up a range of atmospheric humidities in a darkened multiple-choice chamber (see Fig. 7.9 and the graph in Fig. 7.10), Darlington induced a concentration of an optimum number of beetles over a saturated solution of KNO_3 at 25°C which established a relative humidity of 92.5 in the atmosphere above it. When presented with the alternative of moist air in white light and dry air in darkened conditions, the beetles moved to and fro for a considerable time before eventually settling in the humid sector.

Differential distribution of food might operate also, but the catholicity of the beetles' diet precludes expectation of this being decisive. Experiments by Darlington using choice chambers confirmed the broad dietetic range: he found apparent preference for living insect larvae and earthworms, and both Paxton and he

141

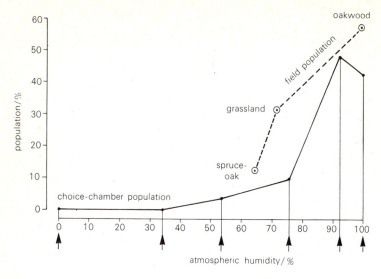

Fig. 7.10 AUTECOLOGY OF *NEBRIA BREVICOLLIS*. Distribution of beetles among humidity variants in the field (Malvern, Worcs.) and in the laboratory. Choice chamber tests were made on 100 insects in 10 groups of 10 each. Correlation between the two sets of data points to humidity as one of the factors affecting this species.

recorded dry food as being taken very little, although it remains uncertain whether this denoted avoidance of unsuitable food material or that of a dry environment. Extraction by Tullgren and Baermann funnels (see Fig. 8.1) from top soil at the various situations were found during Darlington's investigations to yield more worms and larvae in grassland and the oak stand than in the spruce oak plantation – seemingly this accords with the spatial distribution of the beetles. It is, perhaps, noteworthy that earthworms tend to desiccate quickly in dry atmospheres.

On the assumption, therefore, that grassland to a coniferous/deciduous situation and thence to deciduous woodland represent stages in succession towards climatic stability, the expectation would be for *N. brevicollis* to fall away from a relatively high population in one of the early phases and eventually to reach a peak in the highly energized climax. At the same time, the concept of succession as a series of discrete stages, each developing simultaneously over wide areas, is obviously as untenable as the view that human societies develop simultaneously. The relative positions of the situations are important in that a particular stage in a particular locality can affect the success of organisms occurring in places close by.

8 Collecting methods and sampling populations

8.1 Methods of collecting

Information on the size of a population of animals or plants cannot be obtained by direct observation alone, hence some means of collecting must be adopted. Before we describe *how* numbers can be estimated we must outline some of the methods that can be used in the case of animals to collect or to extract them.

Collecting involves the use of mechanical methods combined with a knowledge of the behaviour of the animal species in question enabling the use of appropriate forms of traps.

A brief outline only will be given here of some of the more usual methods, for a more detailed account of these and other methods the reader is referred to the various specialist books and papers mentioned in this chapter.

Examples of the mechanical means of collecting include various types of net such as mist nets used for trapping birds, sweep nets for the collection of insects from herbage as well as nets of a stouter nature for sweeping areas of fresh water.

There are, however, practical difficulties involved in net-sampling. For instance there is the ever-present uncertainty of how much material the net has entrapped and hence whether the numbers so caught are representative of the density of the species in the area

---

Fig. 8.1 TWO TYPES OF FUNNEL FOR THE EXTRACTION OF ORGANISMS FROM SOIL OR LITTER: I Basic form of Berlese or Tullgren funnel used in dry-funnel extraction. It is essential that the funnel should be large, steep-sided and smooth to prevent debris and organisms accumulating on the sides. The use of a long funnel will also cut down the amount of moisture condensation on the sides in which small animals become trapped. The litter sample should be in a steep pile or in a block in the centre of the gauze to allow a current of cool air to circulate, thereby avoiding too rapid warming of the sample and allowing the animals to escape before they dry up.

II Basic form of Baermann funnel used in wet-funnel extraction. The coarse-meshed bag containing the sample should be completely immersed in water, held in the funnel by the attachment of a spring clip to rubber tubing fitted to the stem of the funnel. Samples are taken by releasing the spring clip.

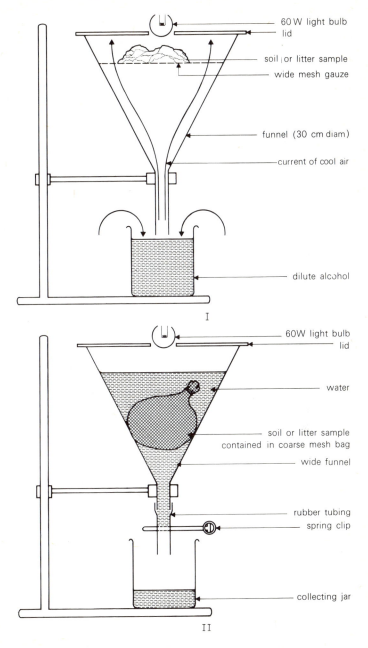

60 W light bulb
lid
soil or litter sample
wide mesh gauze

funnel (30 cm diam)
current of cool air

dilute alcohol

I

60W light bulb
lid
water

soil or litter sample
contained in coarse mesh bag
wide funnel

rubber tubing
spring clip

collecting jar

II

investigated. Again, there may be escapes from the mouth of the net or through its meshes as can happen in the case of ephemeropterous nymphs which often in flowing water and with the aid of the current, pass through even fine mesh. Larger nymphs can escape through the net mouth more readily than the smaller nymphs and this at once imposes a bias when figures involving age-distribution come to be assessed. Various methods for sampling organisms in fresh water are given by Leadley Brown[41].

Sampling by means of sweep nets used on land herbage is frequently inaccurate except for certain species (see sections 6.4 and 8.3) since many grassland arthropods, under certain conditions, will perform a vertical migration so that the net sweeps made in the upper layers of vegetation will give different results from those made nearer the ground.

A more laborious but more accurate method of collecting insects associated with tree foliage such as aphids and mites, is to collect leaf samples at random, wash off the insects using soap solution and count them. This method, although time-consuming especially where a large number of samples must be counted, can yield surprisingly consistent results.

Extraction of soil organisms can be made by using sieves of different sized mesh or by flotation. Since the first flotation methods were used they have become much more sophisticated involving a number of separate processes. Forms of flotation apparatus are described by Laurence[40], who used the method for extracting fauna from cowpats, and by Kevan[39].

There are also various kinds of funnel apparatus for the extraction of small organisms from soil and plant litter samples. In dry-funnel methods (Berlese and Tullgren funnels) the sample is placed on wire mesh at the top of the funnel, the surface of the sample being heated in some way, thereby causing a temperature gradient through the sample and the organisms to move downwards entering the lower part of the funnel, eventually to be collected in a container of preservative fluid into which they fall (Fig. 8.1). In the wet-funnel methods (Baermann funnels) the sample is contained in a mesh bag of some kind, suspended in the mouth of the funnel which is filled with water. A source of heat above the sample causes the organisms to pass downwards through the mesh container into the water where there is a temperature gradient. They eventually pass down the funnel stem and collect in a container of preservative (Fig. 8.1).

This latter method is particularly suitable for the extraction of aquatic organisms in the soil such as Nematoda, Enchytraeidae and Rotifera.

Both funnel methods have been the subject of modifications and where complete extraction and counts of sample populations are required, these modifications have become sophisticated, some involving batteries of funnels so that extraction from several samples can be made simultaneously. Methods of funnel extraction are described by Macfadyen[46], Murphy[53], Wallwork[81] and others.

Apart from the collecting techniques briefly described above, there are other means of obtaining fauna which rely on the principle of attracting the organisms in various ways from their habitat to a point at which they can more easily be collected. To this category belong such things as pitfall traps. These can be a series of glass jars, containing preservative, sunk into the ground in different places. If they are left unbaited, mostly predatory forms are collected. Baiting with attractive substances such as decaying meat, melon or oatmeal is, according to Walker[74], more selective in the fauna extracted. Repellents, like formalin in a 2 per cent solution, or a dilute solution of potassium permanganate (1.6 gdm^{-3} of water), when watered onto 1 m^2 of soil, turf or onto a gravel path, extract earthworms. Collection of slugs and snails can be made using poison bait but such methods when applied to those animals which can move from one locality to another, may result in difficulty in relating the numbers collected to the density of the natural population. First, the bait may attract animals far removed from the actual area. This can be overcome by using some form of grid sampling and marking. Secondly, the susceptibility to the bait may vary with age and with the diurnal or seasonal habits of the animal. In this case successive sampling must be undertaken and results interpreted using a knowledge of the behaviour and phenology of the species.

Emergence traps used to collect insects emerging from immature aquatic forms in streams are described by Mundie[50].

Live traps, such as the Longworth and Tring traps, can be used for small mammals. These rely on bait. Although such traps are useful for the collection of specimens for marking or for other observations, it has been shown, using smoked paper at the entrance, that less than fifty per cent of animals approaching the trap actually enter it. Hence, as a means of sampling a species, such methods on their own, are not accurate.

147

Sometimes larger animals such as nocturnal birds and mammals, are difficult to find or trap. In these cases it is often possible to use their traces. Pellet counts were made by Elton and others[24] for estimating populations of voles and this procedure has been adapted for investigating rabbit populations. Owl pellets can similarly be used.

8.2 Some problems of sampling

In the case of static plants, actual counts of numbers within a community may be possible and is obviously an easier process than counting an animal species which may be dispersed over a wide area and also be on the move from one place to another within a habitat. Even with plants, a complete count is often impossible with small species.

For the physicist isolation of factors, comparatively easily, enables the accurate assessment of the affects of a particular factor over a period of time, but the biologist is usually faced with a number operating simultaneously, giving rise to errors which nevertheless may be measurable. It is evident that the more factors that are involved the more complicated the analysis of results becomes and the greater are the number of samples needed in order to give a required degree of accuracy.

Much time can be consumed if estimation of numbers in every case depends upon an accurate and complete count. By employing various sampling methods, however, it is possible to gain a sufficiently good idea of the distribution and numbers of a population provided that various important points are decided upon beforehand. The time involved must be assessed against the degree of accuracy required and whether actual or comparative results are needed. Again, consideration must be given to the experimental design of counting procedures which will depend on the species to be sampled and the area in which they ae sampled. The positions at which samples samples are taken must be free from personal choice and done without bias in any way.

Samples can be distributed along a transect, but this may lead to errors since certain species may follow physical gradients. Hence the procedure which is decided upon should relate to whether distribution is uniform or not.

In some cases, species will form clumps around areas most suited to their requirements. For instance crane fly larvae in grassland are often found in groups of as many as thirty individuals surrounded by areas in which there are none, distribution probably being influenced by soil humidity or the absence of grass roots or both.

On the other hand, certain species of orabatid mites are usually fairly evenly distributed in the soil in which they occur. Many examples of clumping come to mind. They may be due to physical factors or, as in the case of many kinds of insects to social tendencies to congregate in one place. The black garden ant, *Formica nigra*, nests under stones or turf in large numbers, while several thousand individuals of *F. rufa* are present in their large mound nests. The numerous species of ephemeropterous nymphs to be found in flowing water distribute themselves according to current speed, the type of substrate and the degree of darkness afforded by stones in the bed of the stream. All these factors and many others can play a part in the clumping of species and may require special attention in planning a programme of sampling.

In all cases of non-uniform distribution where populations are grouped with the groups distributed at random, calculation is based on the *negative binomial distribution* derived from the Poisson series. Statistical and mathematical models are described in various texts and papers such as Snedecor[72], Anscombe[2], Brian[7], Southwood[73], Thomas[78], Bliss & Owen[6], and Bishop[5].

8.3 Estimating numbers

When faced with populations of moving organisms, whether they move about in the soil, on land, in water or in the air, various methods of estimating the size of a population can be used. The problem with all methods of sampling is not only to distribute the samples completely at random but also to achieve a compromise between the degree of accuracy required and an adequate total sample size. This problem will be present whatever system of sampling is decided upon and it may well be that more than one method should be used in order to test validity. Three basic methods are discussed here:

(1) *Total counts from areas of known size*
Quadrats are usually employed for sampling limited areas. These are small areas of a certain size bounded by making a wooden or

light metal frame say 10 cm or 1 m square according to the organism or organisms to be counted and within which a total count is made. Sometimes the quadrat is best made as a box trap from which such animals as spiders or insects found in field vegetation can be extracted. For estimating numbers of earthworms in an area, an extraction within a simple frame quadrat 1 m square is made by watering a 2 per cent formalin solution within the quadrat. Complete extraction may require a second watering of formalin after a lapse of 20 minutes. Further samples are taken at various random points within the area of investigation.

(2)   Removal Sampling

This method involves the removal from a known area of successive samples using a sweep net in the case, for instance, of insects in herbage or using a pond net for insects or other small organisms from water. The same number of sweep strokes are used for each sample taken within the demarcated area. The organisms in each sample

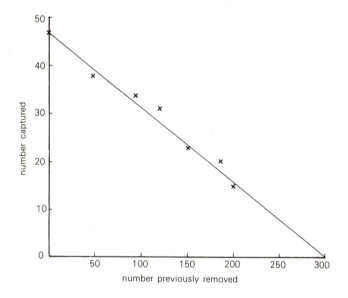

Fig. 8.2 REMOVAL SAMPLING OF A POPULATION OF LEAF HOPPERS IN 10m² OF GRASSLAND BY SUCCESSIVE NET-SWEEPS. The number captured are plotted against the number previously removed. The grouping of points are plotted around the straight line proves that in this case removal sampling was a valid method for estimating the population of leaf-hoppers.

are then counted, labelled and stored separately. Theoretically as more and more individuals are removed, the numbers obtained in successive samples will decline. In an actual case employing the removal system, successive sweeps were made in an area of grassland for leaf-hoppers, *Jassidae*. Numbers captured were plotted against numbers previously removed (Fig. 8.2) for the seven samples taken. The grouping of points around the straight line indicates that probability of capture remained constant and if the line is projected to the zero point, the figure of approximately 300 individuals for a total removal is obtained. In this case, therefore, the adoption of removal sampling proved to be valid. In other cases it might well not be the case, although worth trying out as a test.

Using a net for sampling in this manner may often yield results which when graphed show points which do not fall in a straight line, indicating that successive samples are not constant. Grasshoppers are a case in point, for after the first sweep of an area, the number falls very rapidly indicating that the sweeping operation has, in fact, caused them to leave the area. Many grassland species show a vertical migration in response to changes in humidity, atmospheric pressure, diurnal fluctuations and so forth, so that these physical and biotic influences must be taken into consideration.

(3) *Marking and recapture*
Using this system, individuals from a sample of animals (n) captured from a designated area are marked and released into the population from which they were removed. After they have been allowed to mix randomly within the population, a second sample (s) is taken and the numbers of marked individuals (r) determined. The total population (p) from which the samples have been removed can be calculated using the following formula:

$$p = n \times \frac{s}{r}$$

This system, sometimes called the *Lincoln Index* method, clearly has its limitations, since its validity is dependent upon:

(a) the ability to mark the animal in such a manner that the mark is both permanent and distinguishable;
(b) the mark having no harmful effect nor acting as a deterrent to the random re-distribution of the marked individuals within the population from which they were withdrawn;
(c) the population at the time of the experiment remaining relatively

static in numbers;
(d) a high ratio of recaptures.

Environmental factors may cause increases or decreases while at the same time the age structure of a population must be taken into account.

While it is relatively easy to hit upon a system of marking which will conform to the requirements mentioned above, the various factors mentioned in (c) can easily influence results and if this is the case statistical methods must be employed to interpret results for which the reader is referred to other works mentioned in the bibliography, such as Orians & Leslie[57], Williams[83], Smith[71], Quenouille[62] and Bailey[4].

Experimental work involving marking and recapture carried out by the authors in two separate localities, are given below. Each investigation covered several years and posed its separate problems but they serve as examples of the application of this method for estimating the size of a population.

8.4 Estimating a population of common toads (*Bufo bufo*)

The methods adopted in making this survey essentially depended upon a knowledge of the habits of the common toad especially during the breeding period which immediately follows hibernation.

The area in which the survey was made included gardens near the centre of Exmouth (population 20 000) and were continued during the breeding season for three consecutive years.

It is common for a population of toads to select a single breeding site and in this case the site was a small artificial pond about 5 m long by 2 m in width. The source of water was rainwater from the roof of the house in the garden of which the pond was situated. This drained via the downfall pipes from the roof into a pipe leading to the pond which had an overflow at one end. At its deepest point the water was 55 cm in depth and during a period of no rain in late winter could fall to 50 cm.

The common toad is an 'explosive' breeder, i.e. all sexually mature individuals migrate from their terrestrial hibernaculae to the breeding site within a few days of each other, the population in one pond coming from a single locality in which they have all hibernated. They are usually accompanied by a number of immature males. The onset of this migration appears to be determined by the temperature of the air and it will be noted from the graphs in Fig. 8.3 that in each of the years 1953–1955 the first arrivals of male toads did not occur until the air temperature was 10°C or more. At such times of mass migration it is common to see large numbers of toads slaughtered on the roads and in Switzerland enlightened road authorities erect special 'toad signs (Fig. 8.4) warning traffic of such migrations in the vicinity. At Exmouth despite the absence of toad signs, no such slaughter was observed in the roads near to the spawning site, which indicated that the population most probably was able to travel through surrounding gardens without having to cross a road.

Male toads can become mature in their third season whereas the females do not usually reach maturity until their fourth year. Migration to the breeding site takes place during the night. This may last for ten days or more although in 1953 practically the whole colony reached the pond in question within 24 hours. There are usually fewer females than males.

During the period of actual migration many enemies such as rats can be encountered. In another locality, the writer has seen many toads killed by herons, their beheaded corpses with skin removed, being evidence of this particular predator.

Being a small pond with clear water and little mud, it was possible to capture and mark all toads in the pond. The numbers appearing in the graph in Fig. 8.3 relate to males only and from these an estimation of the numbers of male toads only was made.

Some permanent method of marking was required if the marks were to be distinguishable in succeeding years. Spotting with paint or the use of colour rings were discarded as unsuitable since the former method would mean total loss of the mark at skin-shedding and the latter has been shown to cause sores. Toe-clipping was the system employed and was not only permanent but entirely harmless to the toad, in no way affecting its survival nor its ability to return the following year or to mix with the population during the non-breeding season. The method also offered, for each individual, innumerable combinations between the eight front digits and ten toes. After capturing and marking all males in the pond they were released again. During the period that the males were in the pond, counts were made nearly every day except when the pond was

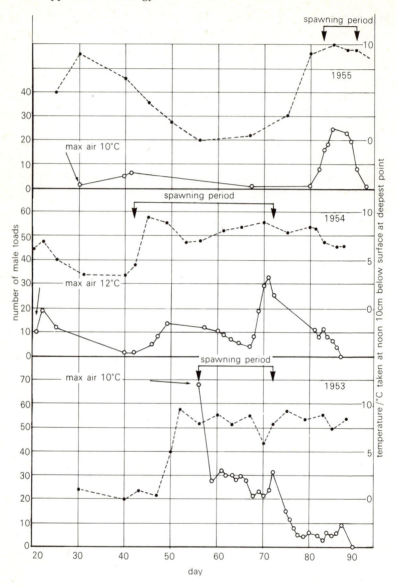

Fig. 8.3 NUMBERS OF MALE TOADS (*Bufo bufo*) IN AN ARTIFICIAL POND DURING THE BREEDING SEASON IN THE YEARS 1953, 1954 and 1955. Numbers of toads are shown as a continuous line. Water temperatures, taken at noon 10 cm below the surface at the deepest point, are shown as a discontinuous line.

Fig. 8.4 SWISS ROAD SIGN. This 'toad' sign, photographed in Switzerland, warns traffic that toads, sometimes in their thousands, cross at this point en route from their hibernaculae to a breeding site. In some places, underpasses have been cut to allow toads to migrate unharmed, a practice which could, with profit, be adopted in this country where wholesale slaughter of toads is not uncommon in certain areas during the breeding season.

frozen over, and fresh arrivals were marked. In 1953 a record number of 69 males arrived in the pond overnight, some had already mounted a female en route.

After spawning the females left the pond almost at once, as is usually the case, but the males remained to mate again with newly arrived females. After the last female had departed, the males remained for a week or more. In 1954 there was an initial, and probably 'false', migration of twenty males as early as January 22, but the temperature of both air and water subsequently fell. No females appeared only one male being visible for several days. Two days later one female appeared and a little spawn was laid, most females arriving early in March, so that there was a prolonged spawning period. Numbers of males increased rapidly during late March in 1955 when the spawning period was brief and the number of males

155

less than in previous years.

Results during the three years 1953–1955 are given in the table below:

	Numbers marked	Numbers recaptured the following year	Total captures	Population
1953	24	14	69	97
1954	21	11	53	101
1955	11	5	27	77

The estimated total population given in the final column was calculated by the Lincoln index, but this assumes for each year, that there was no immigration or emigration, and no fatalities. Obviously were these two factors to be taken into account other statistical methods must be employed. However, for the purpose of this survey the figures obtained showed that it was a comparatively small community which became even smaller during 1954–1955. Indeed further work was undertaken in 1956 which revealed a continued marked decline in numbers.

8.5 Quantitative studies of a blue-tit population overwintering in a town

For estimation of some bird populations, the capture-recapture method is particularly suitable because of ease of marking both with numbered rings and plastic coloured rings–the latter, in combination, allowing identification from afar by its colour code of a bird as a particular individual without the need for close re-examination in the hand. All the blue-tits in the following investigation were doubly ringed, either when they were nestlings or by trapping after they had become free-flying.

Flocking habit

Tits in general tend to remain in solitary pairs during the breeding season but they often form flocks outside it, and when hard weather sets in these parties amalgamate into associations of some size. Members of these flocks enter automatic traps baited with food and here they can be ringed.

In many parts of Britain the two species most readily caught in this way are the blue-tit, *Parus caeruleus*, and great-tit, *P. major*. Six hundred and three great-tits were trapped and ringed as free-flying birds within the market town of Bishop's Stortford, Hertfordshire, during a period of five years: of these, 577 were obtained in winter. In the same period, 817 blue-tits were dealt with in the same way. Many individuals of both species were caught more than once. If the two are compared in respect of the times during winter when they were taken, new birds and retrappings alike, a significant difference in behaviour appears. Comparative tables below are set out as percentages.

Trapping successes including recaptures, during winter months
October 1 – March 31 over five years

	Oct.	Nov.	Dec.	Jan.	Feb.	March	
Great-tit	18	29	25	14	9	5	+ 100% (808)
Blue-tit	4	9	17	31	27	12	+ 100% (1 282)

Table sheet 112

Since the birds enter the traps in search of food, the difference clearly involves a difference in the feeding behaviour of the two, the great-tit either running short of sustenance in the field earlier than the blue-tit or showing a physiological need in late autumn of the kind of food (including animal and vegetable fat) used as bait. There could also be a difference in flocking behaviour, the blue-tit forming its winter flocks later than the great-tit.

Size of overwintering population
Blue-tits are pre-eminently birds of tree and shrub cover and become scarce where there is open ground. A numerical estimate of the population overwintering in a town can be obtained by the capture-recapture method where some idea has been gained of the area covered by the population under study and samples collected of marked birds from scattered parts of the area.

Bishop's Stortford (human population 20 000) has a central core of four square miles incorporating dense urbanization, gardens with the usual lawns, flower-beds and woody plants, and a river course with open water-meadows running north-south and dividing the region into halves (see Fig. 8.5). Automatic traps baited with food and set up near the edges and in the middle of the region took samples of the overwintering birds. There was a certain amount of movement between the trapping stations so that many blue-tits marked at one trap were caught at others. Colour rings were used to observe particularly the behaviour of those individuals which did not re-enter the traps. Information obtained in this way pointed to there being at least two distinct blue-tit populations within Stortford, the river valley and open water-meadows forming the boundary between them. There were indications of a third population in the north-east towards a mixed wood of oak, ash and silver birch.

The midwinter populations of the Stortford blue-tits were compared for two seasons, four years apart: in both, sample *n* was taken in the four weeks January 1–28 and samples *s* and *r* in the period February 4–March 3. Results gave:

first year

$$42 \ (n) \times \frac{29 \ (s)}{6 \ (r)} = 203 \ (p)$$

fifth year

$$51 \ (n) \times \frac{38 \ (s)}{10 \ (r)} = 194 \ (p)$$

The similarity between the two totals, notwithstanding the lapse of time, is extraordinary. It points to a situation of remarkable stability wherein a population of some 200 is maintained, despite the complications of birth, migration and death. At this point it is fitting to stress how misleading a general impression can be even to a practised observer: thus, the late James Fisher, who spent a year in the town, remarked to the writer, 'There must be thousands of blue-tits here'.

Movements

Populations which fall into distinct groups accord with a sedentary species. What movements there are in the blue-tit seem to be circumscribed and determined in range by the need for food. Blue-tits are able to pass freely through timbered gardens; but open grassland or built-up places with very small, treeless gardens are

Fig. 8.5 OVERWINTERING OF BLUE-TIT. Movements during winter months of marked birds between 10 trapping stations in the precincts of a small town. Note separation of discrete populations by built-up terrain and open ground alike. All displaced tits used for homing tests were released outside the area shown. Top–bottom (north–south), 5000 m. Bishop's Stortford, Herts.

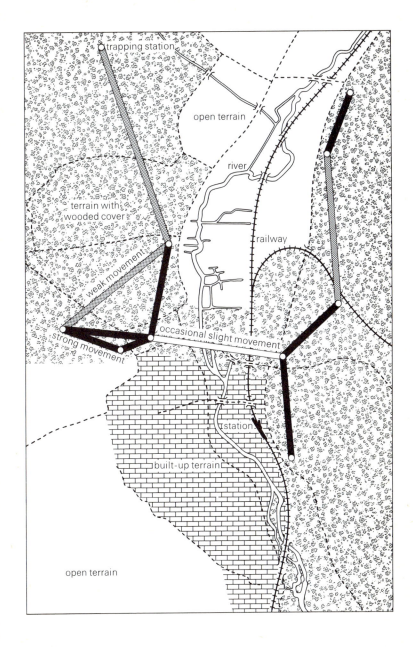

among the natural obstacles to movement. Thin exchanges occurring between traps set up close together but separated by barriers such as these, contrasted with the much stronger movements over greater distances where the gaps were bridged by timbered cover.

In order further to test the impediment imposed by these putative barriers, displacement experiments were tried both inside the town and in the open country beyond. Blue-tits caught at the town traps were released in localities outside their wintering sites and their eventual return to the place where first they were obtained, if it occurred at all, noted from retrap records or identification by colour rings. Displacement releases within the town confirmed the river-valley barrier: no tit was known to pass in either direction east –west across that part of the valley bordered by open grassland, although movement of the kind on a small scale did take place in one locality where there was deciduous shrubbery. Nor did they pass across built-up parts devoid of gardens. A possible complication may have arisen from the large overwintering population in the north-east to which displaced tits released nearby could have become attached permanently if they were lost.

Returns from releases well outside the area of the actual town reflected the same pattern of behaviour. Successful returns came mainly from birds liberated in timbered terrain abutting without interruption on wooded gardens inside the town. Again, one possible factor of importance was the distribution of permanent wintering populations near the places of release which might have 'lured' away some displaced birds.

A question which immediately arises is this–why should these displaced birds return at all? Most of the homing experiments conducted hitherto (e.g. those by Lockley[45] and other workers on the manx shearwater) have been with nesting birds, where the urge to return to offspring might be expected to be a strong one. Provided there is sufficient food in the displacement zone there would seem to be no physiological need for a return to the place of initial capture. One explanation could be that the birds which return most readily are adults (tits over one year old) which are really coming back to their breeding territories. Distinction can be made in the hand between first-winter and adult blue-tits by examination of the plumage of the crown, primaries, secondaries, primary coverts and tail feathers, all of which show minor colour differences: some 80 examinations of the kind tended to confirm the superior homing of

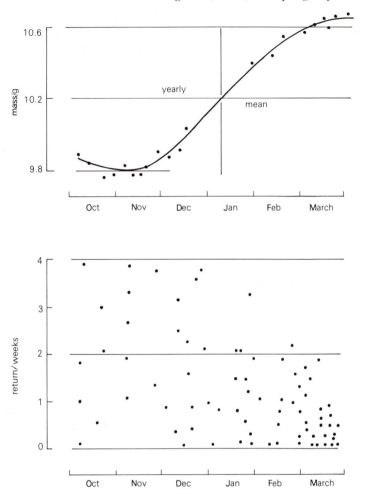

Fig. 8.6 HOMING OF BLUE-TIT. Simultaneous charts correlating increase in body-weight, during maturation of the gonads, with increasing speed of return of displaced birds to original site of capture. Results from two winters. *Upper*, variation in weight from 577 examinations of 482 birds. *Lower*, scatter diagram showing times of first recaptures after displacement from 84 records of 78 individuals: only those retrapped within a month of release are included and all sightings have been omitted. Bishop's Stortford, Herts.

older birds. Additional evidence of another sort was obtained by weighing the birds and noting the time of their earliest sighting back in their original locality after being displaced: as the gonads develop in early spring, subcutaneous fat is laid down. There was correlation between weight and the speed of return: relevant data are charted in Fig. 8.6.

At the same time, it was evident that some of the blue-tits which seemed physically to be identical differed sufficiently in their homing ability to warrant consideration as individuals. (See also Grenville[27] and Hall[30]).

Occasionally, displaced birds lost their way and appeared in unexpected places. A first-winter blue-tit displaced from Stortford to a point five miles to the north apparently overshot the town on its return and was taken in a garden five miles to the south. Another, age unknown, trapped outside the town and released in it, moved to another town 17 miles distant where it fell down a chimney into a living-room fireplace. A third was recovered nearly 30 miles away. To what extent movements inside a wintering population conform to a pattern is difficult to assess, but a cycle of some sort apparently exists: out of 73 blue-tits retrapped in the year following that in which they were marked, 31 were first retrapped within seven days of the anniversary of their original capture and at precisely the same site.

Casualties

There is insufficient evidence for an estimate of the life-expectancy of blue-tits overwintering in this town, although it could be considerable for those birds that had successfully evaded the mishaps which tend particularly to affect juveniles. Thus, there are records of two birds which were first caught and ringed at the end of winter when they must have been at least six months old: both were still coming to traps five years later having, in the meantime, repeatedly returned successfully from displacements of several miles. Casualty-rates among small birds is clearly much higher than the actual number of ringed birds found dead. Carcases of small animals are difficult to locate and are soon removed by saprophages (sexton-beetles, etc.). Reports have come in of only 24 blue-tits being found dead out of 817 marked during the five years: 693 blackbirds of all ages were ringed in the same period, of which 39 have been notified as dead. A comparison of the causes of death is tabulated.

	Predators				Man				Others		
	cat	stoat	owl		shot	traffic	netting	fire		ice	found dead	
Blue-tit	5	1	2		1	1	0	1		3	10	= 24
Blackbird	13	0	1		6	7	5	0		0	7	= 39

Blackbirds feed more on the ground than do blue-tits and this, coupled with the ringing of many nestlings of the blackbird, probably accounts for some of the differences: for instance, the high ratio of blackbirds killed by cats, and injury and death due to entanglement in fruit-netting. Shooting was mainly the work of boys with airguns; and blackbirds are easier targets than blue-tits. A comparison of the times during the year when dead recoveries were obtained also reflects differences due to nestling ringing. Blue-tits surviving their first summer seemed to suffer most casualties in the hard weather of late winter (see table below).

	Jan	Feb	Mar	Apr	May	June	July	Aug	Sept	Oct	Nov	Dec	
Blue-tit	3	7	4	1	0	0	2	2	1	1	1	2	= 24
Blackbird	3	3	4	4	1	7	4	4	2	2	3	2	= 39

8.6 Estimating bird populations by the alighting method

A system of rapidly estimating the populations of certain breeding birds, without the marking (ringing) inherent in the Lincoln index, depends upon their diurnal activity (Vevers & Fisher[79]). The number alighting at a site during a given period of time is compared with the number alighting during the same period at another site nearby (the control), the actual population of which has been measured by direct counting. The ratio between the two groups of alighting birds is that of their respective population sizes. Such a method requires a number of practised observers working simultaneously. Although it was first tried out some 40 years ago, and despite the fact that it calls for sustained concentration, it is probably the quickest and most efficient method yet devised for such species as the gannet and kittiwake which breed on precipitous cliffs and whose colonies are frequently large, diffuse and difficult of access.

Darlington has used it to estimate the sizes of breeding colonies of black-headed gulls and little terns on sand-dunes and coastal flats, where an observer moving between the nests might place the birds at risk. Inland, his students have used it for census work on the rookeries of Essex. Here some of the largest colonies were in woods which were hard to reach, but where rooks alighting at their nests could be counted from afar. Certain of these rookeries, selected later for checking by direct counts of nests from beneath the occupied trees, showed the alighting method to have been accurate to within ± 15 per cent.

The essentials of the system could be adapted for other animals and other situations.

9 Man's place in nature

Man's activities are so varied and on such a vast scale that, in effect, he creates his own environment. Over much of the world, the landscape he sees around him is the result of his workmanship.

9.1 The destruction of climaxes and the effects on soil fertility

Man's destruction of natural climaxes is commonly followed by substitution of other climaxes better suited for his immediate gain whatever may be their long-term consequences for him and for other organisms. Two examples are farms and towns.

Both are so complex that their comparison with natural climaxes may not at once be apparent. Any farm comprises a variety of situations, many of them inherently ephemeral but stabilized for a time by human control. They include the extreme instability of freshly turned soil, which bears little comparison with natural phenomena except such sites as landslides or with the slopes of material ejected by burrowing vertebrates. Changes in a town are those of replacement rather than of succession so that there is a climax situation maintained, in this case, by the multifarious activities of a single animal species instead of by the influence of climate or large plants.

The Neolithic (Late Stone) Age, bringing about its social revolution from predation to agriculture, involved the cultivation and development of annual and biennial plants which could be utilized for crops. These, in turn, necessitated the provision of the special conditions wherein annuals and biennials, the pioneer colonizers of many terrestrial situations, could survive. This meant the destruction of forest climaxes resulting in the exposure of bare soil. Here we are concerned particularly with two consequences of this policy: erosion and diminution in the efficacy of cyclic replacement.

Removal of arborescent cover and the consequent exposure of topsoil to invasion by pioneers may be associated with exposure also to the forces of erosion. Binding of soil particles by plant roots is one form of conservation very difficult to simulate by other means. The particular value in this connection of arborescent root systems must be considered in conjunction with the tendency of trees to accumulate the fall from the atmosphere of precipitation and to direct it into the ground, where they are established. Depth of the roots in soil is less important than their area. Even such large trees as oaks, which may reach a height above the soil surface of 40 metres, have shallow roots spreading laterally but rarely penetrating to a depth exceeding 2 m, so that there is consolidation and retention of topsoil over a wide expanse.

In Britain, where the climate particularly favours the development of grassland, this is frequently a stage in succession after the elimination of forest and can reduce erosion effectively. One cause of failure to retain topsoil against erosion is where over-grazing results in herbaceous shoots being cut so close to the ground that exposure dries the soil which is then readily carried away in wind or runoff, notwithstanding the function of grass roots and rhizomes as binders. Erosion by water where over-grazing prevails is demonstrable by collecting and measuring the soil carried in runoff from turfs grown in trays, irrigated, and differentially clipped. Erosion is greater where the foliage is cut by scissors close to the soil surface than where the leaf-blades are left to develop normally. Results are more significant if cutting and pressure are imposed simultaneously, so as to simulate the combined effects of grazing and trampling, than if the emergent shoots are merely clipped. Many of the grasses appearing in reversion from woodland are shown effectually to provide this sort of cover.

In a near-extreme form, the protective effect of rhizomes is seen in the retention of wind-borne sand by marram-grass in coastal dunes, where progressive increase in the height of the assimilatory shoots and the replacement of the initial rhizomes by fresh lateral growths at higher levels in the substrate keep pace with the gradual piling up of the sand (Fig. 9.1).

Following destruction of forest, the soil is deprived of the kind of recharging which results from leaf-fall and the return of timber debris. Trends in agricultural practice are towards the harvesting of crops before there has been opportunity for an adequate return to be made of organic matter to the soil so that, irrespective of the type of crop or any question of rotation, there is a drift towards a decline in topsoil fertility requiring the addition of chemicals and other materials from outside to offset it.

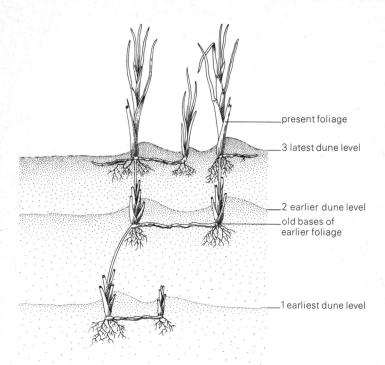

present foliage

3 latest dune level

2 earlier dune level

old bases of
earlier foliage

1 earliest dune level

Fig. 9.1 ECOLOGY OF DUNE-FORMATION. Coastal sandhill, depth 1 m, with wind-borne particles slowed down and accumulating around emergent marram grass (*Ammophila arenaria*). Sand-binding rhizomes and rooted nodes lie in organic matter from earlier levels and the assimilatory growths keep pace with the new surfaces.

In tropical rain forests particularly, it is not so much soil erosion but the loss by leaching from the soil below a cleared forest of nutrients, hitherto localized and circulating within the ecosystem. Nomad agriculturalists have much to answer for in the lowering of the nutrient potential of the land they clear by their practices, and it takes many years to restore the balance. In the Middle East, the Bedouin is not the son of the desert – he is the father of it.

Many the same considerations apply where grassland replaces the original climax, stock is grazed, and the animals are taken away to market before there has been an organic return commensurate with the removal. Formerly it was a common condition of land tenancy to require that excrement from stock should be put back into the ground. Today the situation has been aggravated by changes in husbandry leading to the maintenance, for long periods or even permanently, of animals under indoor conditions when the slurry becomes an embarrassment instead of an asset. But whatever the practice, the situation is one of diminishing return: manure and urine are limited in their long-term value under agricultural conditions and even the ploughing in of entire bodies of dead beasts would merely slow down the accumulation of the debt.

Such considerations have nothing to do with the prevention, by human activities, of access to soil of high potential. One instance may be cited to show how serious this can be. Fitter[25] has pointed out that, with the expansion of London during the past two centuries, some of the most valuable market-garden soil in the country became covered over to provide sites for buildings and roadworks. This is being repeated in other towns.

9.2 Weeds and competition

Inevitably exposure of the soil for agriculture also exposes it to colonization by those weed species which require the same conditions and which may compete vigorously with the cultivated plants. Two advantages in particular accrue from growing a crop as a pure stand–competitors tend to be crowded out and harvesting is facilitated. A serious incidental disadvantage of the practice is that organisms depending on the crop are encouraged by the abundance and proximity of the food supply. This, in turn, induces the employment of chemical pesticides as a defensive measure. Crop rotation, in breaking the life-cycles of pests, is an alternative.

Four general considerations need to be taken into account in this connection. Firstly, although it has become commonplace to decry the use of chemicals as potential promoters of eco-catastrophes, those introduced as herbicides or fungicides tend particularly to be so localized in their application as to leave the surroundings, beyond the limits of the crop stand, relatively unaffected. Secondly, the analytical techniques themselves have become so refined as to allow the detection of the most minute traces of a substance in the environment when the quantities are not necessarily injurious to metabolism. Thirdly, those substances which break down readily under natural conditions are clearly less of a potential threat than the more stable compounds. Finally, if the utilization of pesticides was suspended altogether, there is no doubt that the economy of many countries–notably that of underdeveloped regions–would seriously be affected.

Should injurious organisms become over-abundant, a measure of chemical control in normal circumstances is probably imperative in the absence of effective application of biological controls. This is a subject in its infancy which requires further experimentation under a variety of conditions. Much depends on situations near to the crop. Where these include extensive areas overgrown with weeds, sufficient natural enemies may persist in the cover here to control certain of the pests, and prevent their populations attaining epidemic proportions.

9.3 Relationship of plant cover to erosion

Erosion of topsoil is a product both of human influence and topography.

Parts of Scotland provide an example. Following the failure of the Jacobite rebellion in 1745–6, large tracts of the Highlands were cleared of forest and the successional grassland was exploited by being put down to sheep-grazing. Removal through erosion of the soil beneath over-grazed pasture was enhanced by the incline of the terrain. More than two centuries later the process persists, so that subsoil, and even parent rock deprived of its capping of protective soil, continue intermittently to be washed away after heavy precipitation. Impoverishment of the soil, and the replacement of a timbered situation by moorland or heath, were early developments. (See Fig. 9.2 for an English example.)

Up to the present time, myxomatosis has proved to be the most effective biological control known, and it is a matter for regret that its original role of reducing enormously the rabbit population in Britain during the 1950s was succeeded by wavering, indecisive legislation which prevented administration of the coup-de-grâce and the extermination of this serious pest. But even as matters stand, the rabbit is now subordinate to the sheep and, in some regions, to the goat as an agent of erosion through the destruction of plant cover. Man is responsible for the introduction of all three animals, and visits paid by Darlington over a period of several years to the rabbit-infested island of Great Saltee in the Irish Republic revealed some of the effects. The advent in 1961 of myxomatosis on Saltee was succeeded by a lush sward of dense herbage brilliant with

Fig. 9.2   Fig. 9.3

Fig. 9.2 ECOLOGY OF SOIL EROSION. Buildings threatened through subsoil and parent rock being washed down from high ground by precipitation after excessive destruction of plant cover. Malvern Hills, Worcester–Hereford border. (See erosion in Scotland, page 169.)

Fig. 9.3 ECOLOGY OF SOIL EROSION. Self-sown coltsfoot (*Tussilago farfara*) holding loose soil in steep mound, 6 m high, on top of buried refuse. Covered buds of coltsfoot give rise to new assimilatory growths which elongate into the light with the progressive piling-up of fresh deposits, while its underground structures bind together soil particles. (See marram grass, Fig. 9.1.) Blockley, Gloucs.

Fig. 9.4 ECOLOGY OF SOIL EROSION APPLIED. Strip of coltsfoot, width up to 80 cm, planted along contour of housing estate on hillside to fix soil in bank newly seeded with ryegrass (*Lolium perenne*). Mowing eventually destroyed the coltsfoot and promoted the grass. Malvern, Worcs.

170

flowers whereas, only a year or two earlier, the ground had been carpeted with grasses bitten off close to the surface. Bracken had spread in the wake of the weakened pasture and erosion in the more exposed parts of the island was a persistent feature.

Causes of erosion additional to those mentioned above include the pressures on the countryside like the abrasion of cover which develops in places of concentrated recreational activity, such as sites of antiquarian interest, picnic localities, beauty spots, nature trails, and public rights-of-way. The banks of waterways can be destroyed even by the wash from pleasure boats. Speight[75] deals in some detail with the connection between this problem and the larger issue of conservation.

9.4 Pollution

Pollution includes anything released into the environment which degrades it. In the last two decades, it has become evident that man is responsible for serious degradation of his environment. A chain of events has seemingly brought man to the verge of destroying his own environment, and the first link in this chain is his over-production (Chapter 10). His population explosion, together with a continuing rise in the standard of living, impose on the environment increasing demands for larger and better supplies of food. Economic considerations tend to carry more weight than the efforts of con-servationists to maintain his environment and that of the organisms which share it.

The atmosphere is probably the part of the environment most obviously contaminated. It took a major disaster, the London smog of 1952, to stimulate effective measures against smoke which, although the worst of the atmospheric pollutants, is perhaps the one easiest to control. The smokeless zones which were introduced brought about a considerable improvement. In London, smoke had become reduced by 75 per cent after 15 years. Control of industrial fumes has been less effective: thus, sulphur dioxide remains high and the gas forms acids with rain water which become washed down into the soil.

Inasmuch as there has been an increase in such fungal diseases as mildews, rusts and smuts, smoke abatement can be held actually to have had an adverse effect on plant life.

Combustion of fuel in vehicles has brought its own problems. The lead and carbon dioxide content of the air from exhaust fumes have continued to increase with the continued increase in road traffic. Carbon dioxide occurs naturally in low concentrations in the atmosphere. Its increase is bringing about a 'greenhouse' effect in that it permits penetration to the earth's surface of heat rays from the sun while filtering off some of the others. The general rise in temperature is calculated as being in the order of 0.2°C by 2000 A.D.

9.5 Pesticides

Chemical substances released to destroy invertebrate pests are an environmental degradation of a special kind when they include such materials as the chlorinated hydrocarbons (dieldrin, aldrin, heptachlor, etc.). Sooner or later, pesticides reach the soil. Unlike those of the organo-phosphorus group, the chlorinated hydrocarbons can remain in the soil for an unknown number of years without breaking down, and come to affect organisms outside the range of those against which they are specifically directed. The danger is enhanced through run-off from contaminated soil reaching water-ways.

Recognition of the extent of the problem is comparatively recent. Even in 1963, the Nature Conservancy was expressing doubt as to whether there was sufficient evidence to attribute the decline in the populations of certain freshwater birds to pesticides. In fact, the poisons affect successive levels in a pyramid of numbers (Chapter 5) with increasing severity. In the soil of orchards, sprayed several times a year to control capsid bugs, woolly aphid, codling moth, etc., the build-up of DDT can be fatal to birds feeding on the invertebrates generally. Water birds, like others, tend to lie near the ends of food-chains.

After spraying with DDT, insects and some species of mite in the topmost soil layers are killed off. Thereafter there is an increase in the population of collembolans which are herbivores, caused by the decrease in their natural enemies, the soil mites and predatory insects. The long-term effects of upsetting the balance in the soil may not immediately be apparent and equilibrium may take years to become re-established.

Disturbing the balance may even promote a pest. Where DDT is used, the comparatively slow-breeding insects which normally control red-spider mite are killed off, but the pest itself is resistant to the chemical which, in low concentrations, even stimulates the hatching of its eggs.

Taken further, it requires little imagination to realise that deleterious sprays entering the soil can rapidly be transferred by drainage of the soil into streams and rivers. Filter-feeding organisms, such as the freshwater bivalve molluscs, sieve a large volume of water through their gills from which they extract suspended food. They, and freshwater fish which operate a similar system of extracting oxygen from the water passing over the gills, are the first to suffer. Within their bodies there is a build-up of the chlorinated hydrocarbons which they pass on, in larger amounts when they become the food of water birds such as grebes, coots and moorhens, and of the large predatory fish-eating birds such as the osprey, the species which has most often hit the headlines.

Predators lying at the ends of the chains, accumulate toxic substances with the prey they take so that they either are killed by direct poisoning or their behavioural patterns are affected to their own detriment. Accipitrine birds exemplify predators which are so influenced. Decline in recent years in the numbers of peregrine falcons has been accompanied by changes in behaviour so that the birds have come to destroy their own eggs. Chlorinated hydrocarbons detected in the birds' tissues and eggs have been found by experiment to act as nerve poisons and presumably cause suicidal disturbances of this sort. But it must be recognised that, were DDT for example suddenly to be discontinued, human deaths from insect-carried diseases would greatly increase in under-developed countries. In India alone, and in a single decade, DDT reduced the incidence of malarial cases from 75 million to 5 million so that the curtailment of its use would mean more disease, less food and more human suffering.

Apart from the pesticides as such, persistent effects of the organo-mercury compounds employed as fungicides operate through food-chains. A serious feature is the pollution from paper mills where mercury is employed to control accumulation of slime during manu-facture. Effluents from paper factories can kill by affecting shellfish in estuaries and passing along the food chains to organisms feeding on these molluscs. Nevertheless, a small amount of mercury is a natural feature of the environment, although methyl-mercury compounds are especially dangerous. If it is present, mercury vapour can be synthesised into methyl-mercury compounds by bacteria.

9.6 *Some alternatives to the use of pesticides*

Two alternatives to the indiscriminate application of pesticides, based on ecological relationships, can be summarised as the minimum-kill technique and the use of biological controls.

Minimum kill
Here no attempt is made completely to wipe out the pest by direct poisoning, but its numbers are reduced sufficiently to initiate progressive diminution as a consequence of difficulties in mating, etc. where the population has become small and thinly scattered. There is avoidance of the build-up of strains immune to specific chemicals: the method involves a more selective and careful use of the poisons employed.

Biological controls
Biological control, like a controlled experiment in scientific investigations, is something of an ideal. The precise conditions under which a natural enemy can be employed to reduce effectively the population of a pest have still to be elucidated, although these probably include four of paramount importance.

1 The control must be able to survive the particular climatic conditions in which the pest exists.
2 Its food-range must be narrow so that it concentrates on the organism whose control is desired.
3 It must be able to locate this organism.
4 Its own changes in population must keep step with changes in the population of the prey.

Glasshouses in Britain seem to favour the operation of biological controls against pests introduced from warm climates. One example of such a pest is the greenhouse white-fly, *Trialeurodes vaporariorum*, which probably came from Brazil and is now a nuisance to growers of tomatoes, cucumbers and ornamental plants under glass. Originally controlled by fumigants (hydrocyanic gas or tetra-chlorethane, which were either dangerous or costly) it is now controlled by the minute chalcid, *Encarsia formosa*, which produces its ova parthenogenetically and inserts these into the white-fly nymphs, killing them. Its efficacy as a control is probably aided by the fact that there is no need for mating, thus eliminating the problem of the sexes locating each other. Both the parasite and host require the tropical conditions inside a heated greenhouse for survival from year to year.

Another alien insect introduced into Britain (but which, despite its name, probably did not come from the New World) is a pest of apple orchards, the American blight or woolly aphid, *Eriosoma lanigerum*. Owing to its dense wool-like covering it is difficult to destroy by chemicals after it has hatched; but the discovery in North America of its chalcid parasite, *Aphelinus mali*, has enabled this to be used as a biological control in some countries, leading to greatly reduced infestation in South America, parts of the Australian continent, and New Zealand. However the chalcid itself has not become adapted to British conditions and numerous failures indicate that colonies of the parasite would have to be maintained here under special protection throughout the winter, as is the case with *Encarsia*, so that it could be released among infested trees periodically and in substantial numbers at favourable opportunities.

Britain is not a region wherein to find outstanding achievements in biological control. We could be fortunate in having a changeable climate, cool, relatively sunless, humid and with periodic freezing fog, which inhibits the establishment of organisms from overseas so that they become pests of importance. As far as our indigenous pests are concerned, it is likely that a high degree of biological control is already being carried out by native predators; and if this is so, the introduction of other kinds would prove less successful than might initially be expected. But in different parts of Europe there are different predators on the same individual species of pest. It could be that, in some cases at least, geographical rather than climatic barriers have blocked their spread to Britain. Thus it is possible effective control might be achieved through the introduction of some of these forms from other countries.

Certain of our native parasites have been exported in attempts to reduce those of our insect pests which have accidentally reached other regions. In milder climates than ours, hosts and parasites alike have frequently flourished and there have been notable successes in biological control. The case of the small white butterfly, *Pieris rapae*, in New Zealand may be cited: this arrived in 1929 and multiplied so rapidly and dispersed so widely that, within six or seven years, the invasion of the country was complete and the insect had come to be a major threat to the economy of vegetable-growers. But the introduction of yet another chalcid parasite, *Pteromalus puparum*, which oviposits in the chrysalids and not in the larvae, brought the butterfly rapidly under control and it has ceased to be a serious problem.

Controls which have 'backfired' to become pests themselves include predators which have either been too general in their selection of food or unable to locate in sufficient quantity the prey they were intended to destroy.

Before the advent of myxomatosis, the over-running of the island of Great Saltee by rabbits (see 9.3) induced the island's owner to attempt to control them by releasing cats as predators. Whereas these settled down and became feral, they attacked other animals besides rabbits, including nesting birds which he was seeking to protect. In an effort to control the cats he resorted to foxes which, in turn, proved to be general predators. House sparrows, liberated in the town parks of New York to keep down caterpillars defoliating the trees, certainly destroyed the insects when they were able to locate them, but they were not well adapted for searching out prey high in leafy branches: they moved into open country where they became farmland pests feeding mainly on grain.

Utilization of controls of the *Encarsia*, *Aphelinus* and *Ptermomalus* type has the advantage that they are parasites as well as predators so that they take the narrowest possible range of food.

9.7 *Chemical pollution of water*

The threat of pollution to streams and rivers is one which should be taken seriously. The chalk streams of southern England are unique. Fed by springs of pure alkaline water underlying the chalk strata, they support an abundance of plant and animal life. But the cheapest and easiest way of obtaining water in sufficient quantity to satiate the ever-growing needs of our expanding cities, is to sink boreholes to the sources of such springs, thereby depriving a river of spring water throughout its length. Treated sewage effluents carry high concentrations of inorganic salts which become enhanced by chemicals draining from the land. Counteraction of the effects of such effluents upon streams and rivers lies in dilution. Clearly, upstream abstraction of water has the reverse effect. It should be of prime importance to the Water Resources Board and the river authorities to see that abstraction takes place near the mouth of a river, at the point of maximum flow. This must, of course, involve purification schemes; but sooner or later such schemes must be implemented if we, in Britain, are to salvage what remains of our freshwater habitats.

Rivers wash the poisons into the sea and on the way, estuarine molluscs like mussels and cockles become affected. Terns and oyster-catchers feed on these molluscs and upon the worms living in the estuarine mud which, in turn, feed upon detritus settling on its surface.

It is a remarkable fact that no part of our oceans is now free from these pollutants. Concern is being expressed for the survival of many fish-eating birds in such remote parts of the world as the Galapagos Islands. Pesticides, which seem to be used with little restraint in South America, pour into the sea where the Humboldt Current sweeps them to the Galapagos. Here the flightless cormorant, Galapagos penguin, lava-gull and other species in danger of extinction, are suffering further reduction in numbers through the effects of poisonous hydrocarbons, which act in much the same way as they have done in the British accipitrines and bring about in-fertility.

In northern waters, reference should be made to such auks as the razorbill, guillemot and puffin, the numbers of which have declined in the last 20 years. Two of the factors in the decline appear to be chemical pollution and oil pollution. More than half a million tons of chemical waste are dumped into the North Sea from the Continent every year; and ship oil, petroleum chemicals, and industrial wastes ranging from china-clay rejects to radioactive effluents, wash into the seas around our coasts (see also Allsop[1] and Mellanby[49]).

Problems of pollution are obviously acute in a small and highly industrialised island such as Britain. At the same time there is the temptation to regard sparsely populated regions like the Arctic as 'fair game', where destructive exploitation is tolerable for the sake of a short-term advantage. But a new sense of responsibility is arising, and people are gradually realising that places, which today are commonly regarded as barren wastes, may tomorrow provide valuable resources.

9.8 *Organisms promoted by man*

In Britain as elsewhere, man is responsible for the introduction and maintenance of aliens, either by accident or design.

Commensals sharing his food include the house mouse, which was almost certainly brought in during Neolithic times, and the brown rat, a ship-borne immigrant of the eighteenth century. Human habitations, warm and providing shelter, enable aliens to survive which normally require semi-tropical conditions. These include such forms as the silverfish and firebrat of kitchens and boiler-houses, and the crickets and cockroaches of warehouses and bakeries, some of which can persist for a time in certain outdoor situations like refuse dumps where heat generated by putrefying matter sets up microclimates so favourable to them that they even breed. Waste from man's settlements can operate to assist the success of an alien in competition with an indigenous species: thus the rainbow trout, *Salmo irideus*, deliberately introduced here from North America in 1882–4, flourished in a Derbyshire stream polluted by sewage which killed off the native brown trout, *S. trutta*, thereby allowing the rainbow to take over the vacated stretches of water.

There is always the risk of an alien, deprived of factors which normally keep it under control, multiplying in its new environment to such an extent as to become a pest. In recent times this has happened with the grey squirrel. For a while it was suspected of being true also of the little owl, following its introduction from Holland and elsewhere in 1843–1889 (Witherby *et al*[84]). Its phenomenal increase in numbers and range led to the suspicion of its being a significant predator of game birds and native passerines, but an inquiry conducted in 1936–7 by Hibbert-Ware[36] and based largely on an analysis of its pellets collected from widely separated districts showed it to be, in fact, predominantly a feeder on insects and small mammals.

Human activity has unintentionally brought about the establishment and spread of many alien plants: Canadian pondweed, *Elodea canadensis*, came into the midlands before 1847 with timber from overseas and subsequently was spread from Market Harborough along the canal system by barges and their tow-ropes which fragmented the shoots; and hoary cress, *Cardaria draba*, reached the Isle of Thanet from the continent in straw litter during one of the campaigns against Napoleon and has since spread widely by fragmentation caused through ploughing. Compositae afford some of the most striking examples of the extension, from small initial areas of colonization, by alien species–aided both by transport and by the development in such waste places as refuse tips of conditions favourable to them–which are becoming firmly established over increasingly wide areas. Thus, Oxford ragwort, *Senecio squalidus*, originally an introduction into the botanic garden at Oxford from the ash-covered slopes of the Italian volcanoes, spread with the cinders cast out in household refuse, moved along railways during the steam age when locomotives dropped ashes from their fireboxes, and became well established in towns where incendiary bombs during the last war provided conditions resembling those in its native environment. Rayless chamomile, *Matricaria matricarioides*, a native of Asia which reached us by way of Oregon, was restricted a century ago to a single locality in North Wales from which it has since been transported on the tyres of motor vehicles to waste ground in virtually every district of mainland Britain.

Where the pioneers comprise a small stock there is a particular chance of alien colonizers attaining a high peak and then declining: both the little owl and Canadian pondweed have fallen away substantially in recent years.

Man's constructional activities result in the creation of innumerable artifacts providing situations for exploitation by organisms, both indigenous and exotic. These range from the nail-holes in mortar-courses and the cavities inside broken equipment discarded at refuse sites occupied by solitary bees and wasps (Darlington[18]), to the moss cushions growing on building rubble and stone walls wherein dwell such adaptable animals as tardigrades (water-bears). Small animals from mosses can be extracted by the use of funnels (Chapter 8). It could be that the establishment of mosses on walls, buildings and roofs has increased the frequency and the range of some of these forms. Almost certainly, the inclusion of lime in the binding material of walls has promoted the spread into non-basic regions of certain calcicolous plants capable of withstanding the intermittent aridity of these structures.

10 Conservation or preservation?

Conservation and preservation are not synonymous.

Conservation is the maintenance of the environment wherein organisms can sustain themselves. It must in some degree be selective for the retention and development of 'desirable' amenities. For instance, control of predators of certain species may be imperative if survival of the prey is the particular aim of the programme. Selectivity of this kind makes the distinction between the reserves set up in national parks on the one hand and, on the other, relict areas left to undergo unchannelled development.

Preservation is the maintenance of organisms in zoological and botanical gardens and their equivalents. An aquarium, being a container for a small, selected range of plants and animals requiring constant human attention to survive, is a familiar example of one such situation. In effect it is a water garden and bears the same kind of artificiality relative to field conditions as does an arboretum or an ornamental flowerbed (see also 7.1). A so-called balanced aquarium, with self-perpetuating organisms under minimal control by man, is exceptional.

Preservation and conservation become linked when a species in danger of extinction is maintained under close management, so that its breeding is promoted, and the population reared under the special conditions used to replenish depleted habitats after the factors supposedly causing reduction have been brought under control. Illustrative of this kind of work is the particularly successful case of the Hawaiian goose or ne-ne (Fig. 10.1). By 1950, the world population had become reduced to some 40 birds occupying hollows in volcanic terrain on Hawaii. In this year, three adults were sent to the Gloucestershire station of the Wildfowl Trust where small enclosures were set up in grassland as rearing pens. Within six years, there were 37 ne-nes alive in Europe, and at the Trust grounds every female old enough to do so was laying: in a single season there were 62 eggs of which 22 were fertile; 16 goslings hatched, 15 survived, and three more were produced by a pair on loan to another centre nearby. Twenty years after the scheme for preservation began, 56

Fig. 10.1 PRESERVATION. *Left*, ne-ne (Hawaii), saved when on verge of extinction and re-introduced into its native volcanic terrain. Probably because few individuals comprised the gene pool for replenishing the population, abnormalities have started to appear among their descendants. (Page 180) *Right*, whooping crane (N. America), less than 50 of which survive and a hope for which might be similar methods of preservation.

geese, the sixth and largest single consignment from the Trust, were returned to Hawaii, bringing the total number of re-introductions to 200. The British-reared ne-nes have been set free in a reserve on Maui where the predators, which were certainly an important factor in the initial decline to the verge of obliteration, are now controlled.

Without a link of this kind, preservation for its own sake could result merely in museum situations, where species are kept alive beyond the point at which the natural processes of selection would have eliminated them automatically during the passage of evolution. For example, it is hard to believe that a flightless bird like the great auk (Fig 10.3) could, or should, survive under modern conditions in northern latitudes, even in environments where there is a strong concern for preservation. It became extinct in about 1840 and the depredations of seamen are commonly held to be responsible; but there is more to it than this. It had become a swimmer against the favourable evolutionary tide and, ultimately, out-of-date, and its retention would have been that of an anachronism without faunistic significance.

Fig. 10.2 UNAPPLIED ECOLOGY. Small selection from the many bird-nests deserted along a single hedgerow after cover had been trimmed while breeding was in progress. Species included song-thrush, greenfinch, goldfinch and bullfinch: a brood of the last (centre) is dead. Leigh Sinton, Worcs. (Photograph, *Malvern Gazette*).

Retention of relict situations is commonly put forward as one way of practising conservation. British farmers are recommended to retain 1–2 per cent of their land in the form of hedgerows and so provide cover for organisms from cleared woodlands. This is of limited value if the relicts are isolated one from another: connecting hedgerows are essential for slow-moving animals (the cryptozoic invertebrates, for example) which need to move from place to place with shifting conditions. Far-ranging vertebrates like birds are affected less. And if trees are planted, so nice is the natural balance between organisms that measures designed in the conservation programme may be imperilled by the planting: an instance being the inclusion here and there of atypical trees, such as fast-growing softwoods, as screens for the slower hardwoods during the early

30 cm

Fig. 10.3 EXTINCTION. Great auk (N. European islands), difficult of preservation in commercial latitudes, which became extinct in the 1840s.

stages of development.

Practically everything put forward in this chapter and the last stems from the fact that, for so large an animal and one exercising so decisive an influence, the population explosion which man is undergoing represents the most formidable development of its kind that the world has seen.

In terms of simple numbers its progression is extraordinary. Man took centuries to reach a world population of one billion; thereafter only 80 years elapsed before the population doubled; in another 41 years it had reached 3.7 billion; by the year 2000 it is likely to be 7 billion and by the year 2050 there could be as many as 30 billion people jostling for living space in what had, at one time, been green and productive surroundings. Any likelihood of the various technical processes, necessary simply for mere survival in such conditions, keeping pace sufficiently with developments even to delay the fulfilment of Thomas Malthus' gloomy prophecies, receives negligible

Fig. 10.4   Fig. 10.5

Fig. 10.4 ECOLOGY IN THE SMALL GARDEN. Slow changes. Granite crown of brick pier planted on V.E. Day (May 8 1945) with a single rosette of houseleek (*Sempervivum tectorum*), photographed 29 years later. No human control was applied in the interval. Successional changes have been slight, but the original species has increased vegetatively to become massed into a heap from which rosettes continually fall. Ecologically the site resembles upland rocks where control is less by biotic than by edaphic and climatic factors. Heap, 55 cm across. Malvern, Worcs.

Fig. 10.5 ECOLOGY IN THE SMALL GARDEN. Rapid changes. Narrow ornamental pool, 187 cm across, photographed three years after being planted with a single initial each of greater spearwort (*Ranunculus lingua*) and water mint (*Mentha aquatica*). Both species creep through mud and the changes followed the uncleared accumulation of wind-carried leaf litter from terrestrial plants. Ecologically the pool compares with a larger expanse of water undergoing silting-up. (See also Fig. 6.1). Malvern, Worcs.

credence save from the most optimistic of the constitutionally optimistic.

Clearly regional distribution is more significant than density overall. Thus, the United States of America has 58 people for every square-mile of territory, a figure which does not immediately suggest overcrowding–in fact, it is only one-sixth of that of Switzerland which, in turn, does not appear to have a particularly large population–but 70 per cent of all Americans are jammed together into two per cent of the land.

Initially, some effects of this jamming into a confined space are obvious. There is the problem of the effective disposal of wastes of all kinds. Technical developments may help to solve the problem but they also add to it. For instance, of the 16 million tons of solid waste collected in 1969 in England alone, 256 000 tons consisted of plastics: and the plastic-producing industry is likely to show an annual increase of at least 10 per cent and, moreover, new processes have been developed to manufacture heat-resistant and shock-supporting plastics. Also there is the problem of acquisition of such basic needs as adequate supplies of clean water. In 1970, at least 50 schemes for new housing in Britain were set aside because of shortage of water. Local authorities were warned by the Water Resources Board that difficulties would arise within a few years in supplying even existing homes. But techniques are being developed in attempts to cope with the situation: Europe's largest water-purification plant, at Dömagen in the Ruhr, cleans daily 65 000 m³ of water from industrial installations to within 95 per cent of purity. And the traffic emanating from densely populated regions presents problems additional to those arising directly from the residual gases of combustion. Lead pollution has been traced to the lead alkyls added to petrol and Professor Summer Kalman of Stanford warns that lead from car exhausts accumulates on roadside grasses to produce dangerously high levels of lead in the meat of cattle fed on hay processed from vegetation grown as far as 150 m from the highway. He believes that tests will show traces of lead poisoning in human beings from eating such beef. Furthermore, the nine million tons of de-icing salts used in 1970 to make American road surfaces safe for traffic in snow-bound conditions have been identified as a source of sodium pollution for marginal plant-life as well as for water supplies, and the city of Burlington, Mass., has banned the use of such salts after detecting an excessively high sodium level in its drinking water.

More speculative than any of the foregoing are the behavioural effects of overcrowding and their long-term influence on society itself (see Hall[30]). Whether evidence from experiments on other mammals should be interpreted broadly so as to apply to human societies remains an open question.

So far the most disturbing results have come from the experiments of Dr. John Calhoun of the National Institute of Mental Health at Washington DC (in Taylor[77]), who used laboratory mice. He introduced eight mice as four pairs into an enclosure eight-and-a-half feet square subdivided into 16 compartments. They had access to plentiful supplies of food and water, pollution was removed and predators eliminated. As the colony increased through breeding, they sorted themselves out until 10 were occupying each compartment, i.e. 160 members of a kind of mouse-city, living a well-ordered, busy existence in an affluent society. But in a little over two years the population had reached 2 200 and the first sign appeared of breakdown of the overcrowded society.

This took the form of rejection by adult mice of large numbers of young which then began fighting among themselves, becoming over-active and violent and tearing at their fellows which had lost the capacity to flee. Soon aggression and sexual activity faded and the mice matured into 'passive blobs of protoplasm, physically healthy but frozen in a child-like trance'. The last animal born in the society (it was stillborn, incidentally) appeared towards the end of 1970 and the last survivor was expected to die midway through 1972; but any social structure had already been destroyed (see also Chapter 6.1).

Arising from the problem in the context of human society is the somewhat quaint suggestion that the excessively polite, ceremonial traditions of behaviour of the Japanese are really a response to a long experience of high populations with their concomitant of over-crowding, Japan having already had densities of around 210 persons per square mile over a century ago. Just as territorial disputes between certain birds, such as robins, are commonly resolved by ritualistic posturing without serious injury to the contestants, it might be inferred that Japanese formalism releases stresses of an equivalent kind. As far as a purely physical solution to the problem is concerned, over-population can be reduced, for a time at least, by the intelligent use of land hitherto avoided. How a desert can be made habitable and productive by a combination of technical skills and sheer hard work has been demonstrated in the Negev by the Israelis during the last quarter of a century.

An American biologist, Paul Ehrlich, has expressed the problem succinctly, 'If we don't do something about population and environment and do it immediately, there is just no hope that civilization will persist'. He has the courage of his convictions: he has had himself sterilized.

Throughout much of this book we have inferred the importance of control in ecological relationships. What is controlling man?

Bibliography

1. ALLSOP K. (1970). *Connexions*. Penguin Education.
2. ANSCOMBE F. J. (1950). Sampling theory of the negative binomial and logarithmic series distributions. *Biometrika 37*, 359–82.
3. ASHBY M. (1961). *Introduction to plant ecology*. Macmillan.
4. BAILEY N. T. J. (1952). Improvements in the interpretation of recapture data. *J. Anim. Ecology 21*, 120–7.
5. BISHOP O. N. (1971). *Statistics for biology*. Longman.
6. BLISS C. I. & OWEN A. R. G. (1958). Negative bionomial distributions with a common k. *Biometrika 45*, 37–8.
7. BRIAN M. V. (1953). Species frequencies in random samples of animal populations. *J. Anim. Ecology 22*, 57–64.
8. CHAMPNESS S. S. & MORRIS K. (1948). The population of buried viable seeds in relation to contrasting pasture and soil types. *J. Ecology 36*, 149–73.
9. CLAPHAM A. R., TUTIN T. G. & WARBURG E. F. (1962). *Flora of the British Isles*. Cambridge.
10. CLOUDSLEY-THOMPSON J. L. (1952). Studies in diurnal rhythms. II – Changes in the physiological responses of the woodlouse *Oniscus asellus* to environmental stimuli. *J. expl. Biol. 29*, 295–303.
11. CLOUDSLEY-THOMPSON J. L. (1958). *Spiders, scorpions, centipedes and mites*. Pergamon.
12. CLOUDSLEY-THOMPSON J. L. & SANKEY J. (1961). *Land invertebrates*. Methuen.
13. COLEMAN R. (1972). Fish farming in Israel. *Country Life 152*, 3937.
14. COLLINGE W. E. (1927). *The food of some British wild birds*. York.
15. CROLL N. A. (1966). *Ecology of parasites*. Heinemann.
16. DARLING F. FRASER (1937). *A herd of red deer*. Oxford.
17. DARLINGTON A. (1975). *The pocket encyclopaedia of plant galls*. Blandford.
18. DARLINGTON A. (1969). *Ecology of refuse tips*. Heinemann.
19. DUNCAN U. K. (1959). *A guide to the study of lichens*. Buncle, Arbroath.
20. EDNEY E. B. (1953). The body temperature of woodlice, *J. Expl. Biol. 28*, 271–80.

21. EDNEY E. B. (1954). Woodlice and the land habitat. *Biol. Reviews* *29*, 185–219.
22. EIBL-EIBESFELDT I. (1962). New observations on the marine lizard *Amblyrhincus cristatus. Senckenbergiana Biologica* 30.5.62.
23. ELTON C. (1927). *Animal ecology.* Methuen.
24. ELTON C., DAVIS D. H. S. & FINDLAY G. M. (1935). An epidemic among voles (*Microtus agrestis*) in the Scottish border in the spring of 1934. *J. Anim. Ecology 4*, 277–88.
25. FITTER R. S. R. (1945). *London's natural history.* Collins.
26. GAUSE G. F. (1936). Principles of biocenology. *Quart. Rev. Biol. 11*, 320–36.
27. GRENVILLE H. W. (1971). *Biology of the individual.* Longman.
28. HALL R. E. (1959). Delayed development of eggs of *Chirocephalus diaphanus* Prévost. *Hydrobiologia XIII*, 1–2.
29. HALL R. E. (1961). On some aspects of the occurrence of *Chiro- cephalus diaphanus* Prévost. *Hydrobiologia XVII*, 3.
30. HALL S. (1966). *The hidden dimension.* Doubleday.
31. HARDY A. (1959). *The open sea: II fish and fisheries.* Collins.
32. HARPER J. L. (1957). The ecological significance of dormancy and its importance in weed control. *Seventh Int. Conf. for Plant Protection.*
33. HARPER J. L. (1958). *Ranunculus repens* in biological flora of British Isles. *J. Ecology 45*, 314–25.
34. HARPER J. L. & SAGAR G. R. (1953). Some aspects of the ecology of buttercups in permanent grassland. *Brit. Weed Control Conf.*
35. HEELEY W. (1941). Observations on the life histories of some terrestrial isopods. *Proc. Zoo. Soc. Lond. B III*, 79–149.
36. HIBBERT-WARE A. (1938). *Report of the little owl food inquiry 1936–37.* Brit. Trust for Ornith.
37. JANETSCHEK H. (1949). Tierische Successionen auf hochalpinen Neuland. *Ber. naturw.-med. Ver. Innsbruck 49*, 1–215.
38. KERSHAW K. A. & ALVIN K. L. (1963). *The observer's book of lichens.* Warne.
39. KEVAN D. K. McE. (1962). *Soil animals.* Witherby.
40. LAURENCE B. R. (1954). Larval inhabitants of cowpats. *J. Anim. Ecology 23*, 234–60.
41. LEADLEY BROWN A. (1971). Ecology of fresh water. Heinemann.
42. LEWIS T. & TAYLOR L. R. (1967). *Introduction to experimental ecology.* Academic Press.

43. LIETH H. (1960). Patterns of change within grassland communities. *Brit. Ecol. Soc. Symposium I: Biology of Weeds* (Harper J. L. ed.)

44. LINDEMANN R. L. (1942). The trophic-dynamic aspect of ecology. *Ecology 23*, 399–418.

45. LOCKLEY R. M. (1942). *Shearwaters*. Dent.

46. MACFADYEN A. (1955). A comparison of methods for extracting soil arthropods. *Soil Zoology* (Kevan D. K. McE. ed.) Butterworth.

47. MACFADYEN A. (1963). *Animal Ecology*. Pitman.

48. MACKIE G. O. (1947). The flora and fauna of rain-gutters. *Blundell's Sch. Sci. Mag. 2*, 29–33.

49. MELLANBY K. (1967). *Pesticides and pollution*. Collins.

50. MIDDLETON A. D. & CHITTY H. (1937). The food of adult partridges, *Perdix perdix* and *Alectoris rufa*, in Great Britain. *J. Anim. Ecol. 6*, 322–36.

51. MONTEITH (1973). *Principles of environmental physics*. Arnold.

52. MUNDIE J. H. (1964). A sampler for catching emerging insects and drifting materials in streams. *Limnology & Oceanography 19.3*, 456–59.

53. MURPHY P. W. (1962). *Progress in soil zoology*. Butterworth.

54. NEWELL P. F. (1967). Methods of study in soil ecology. *Paper to* UNESCO-IBP *Paris*.

55. O'CONNOR F. B. (1963). Oxygen consumption and population metabolism of some populations of Enchytraeidae from N. Wales. *Soil organisms* (Docksen J. & Drift J. van der, ed.) 32–48.

56. ODUM H. T. (1957). Trophic structure and productivity of Silver Springs, Florida. *Ecol. Mongr. 27*, 55–112.

57. ORIANS G. H. & LESLIE P. H. (1958). A capture – recapture analysis of a shearwater population. *J. Anim. Ecology 27*, 71–84.

58. PERRING F. H. & WALTERS S. M. (1962). *Atlas of the British flora*. B.S.B.I. & Nelson.

59. PHILLIPS I. D. J. Apical dominance. *Physiology of plant growth and development*. Wilkins M. B. ed. McGraw Hill 1969. Chapter 5.

60. PHILLIPSON J. (1966). *Ecological energetics*. Arnold.

61. PHILLIPSON J. & WATSON J. (1965). Respiratory metabolism of the terrestrial isopod *Oniscus asellus*. *Oikos 16*, 78–87.

62. QUENOUILLE M. H. (1950). *Introducing statistics*. Butterworth.

63. RAUNKIAER C. (1934). *The life forms of plants and statistical plant geography*. Oxford.

64. RICHARDSON J. A. (1964). *Physics in botany*. Pitman. p 74.

65. SAGAR G. R. & RAWSON H. M. (1965). The biology of *Cirsium arvense* (L.) Scop. *Seventh British weed control convention.*

66. SALISBURY E. J. (1942). *The reproductive capacity of plants*. Bell.

67. SALISBURY E. J. (1961). *Weeds and aliens*. Collins.

68. SHEARER G. D. (1938). *Vet. J. 94*, 22–32.

69. SINKER C. A. (1972). A simple experiment in grassland ecology. *Handbook for field work*, 25–26. Griffin & George.

70. SLADE R. E. (1951). *J. Roy. Soc. Arts 94*, 4847.

71. SMITH C. A. B. (1954). *Biomathematics: the principles of mathematics for students of biological science*. Griffin.

72. SNEDECOR G. W. (1949). *Statistical methods*. Iowa.

73. SOUTHWOOD T. R. E. (1966). *Ecological methods*. Chapman & Hall.

74. SOWERBY J. & J. E. (1913). *English botany*. Bell.

75. SPEIGHT M. C. D. (1973). Outdoor recreation and its ecological effects. *Discussion papers in conservation, 4*. University Coll. London.

76. SUTTON S. L. (1972). *Woodlice*. Ginn.

77. TAYLOR G. R. (1970). *The doomsday book*. Thames & Hudson.

78. THOMAS M. (1949). A generalization of Poisson's binomial limit for use in ecology. *Biometrika 36*, 18–25.

79. VEVERS H. G. & FISHER J. (1936). A census of gannets on Ailsa Craig, with a new method of estimating breeding-cliff populations. *J. Anim. Ecol. 5*, 246–51.

80. WALKER T. J. (1957). Ecological studies of the arthropods associated with certain decaying materials in four habitats. *Ecology 38*, 262–76.

81. WALLWORK J. A. (1970). *Ecology of soil animals*. McGraw-Hill.

82. WATERHOUSE F. L. (1966). Alternative or choice chambers for study of arthropod behaviour. *Sch. Sci. Rev. 163*, 726–36.

83. WILLIAMS C. B. (1947). The logarithmic series and its application to biological problems. *J. Ecology 34*, 253–72.

84. WITHERBY H. F., JOURDAIN F. C. R., TICEHURST N. F. & TUCKER B. W. (1938). *The handbook of British birds*. Witherby.

Glossary

Definitions of scientific terms
as used in this book,
a, adjective   n, noun

accipitrine n.
Hawk-like predator.
achene n.
Fruit which does not split
open, containing one seed.
adventitious a.
Structures which arise from
an unusual position.
agamic a.
A generation which
reproduces without sexual
fusion.
A horizon n.
Topsoil and its associated
surface litter.
antibiotic a & n.
Inhibiting life.
apical a.
At or of the growing tip
(apex) of a plant stem.
ascospore n.
Spore produced in a sac
(ascus) by certain fungi.
autecology n.
Study of a single species in
relation to its environment.
autointoxication n.
Reabsorption of poisons
produced by the body.
autotrophic a.
Organisms synthesising
foodstuffs from inorganic
substances.
auxin n.
Growth-regulating hormone
in a plant.
axillary a.
A structure arising in the
angle (axil) between a leaf
and a stem.

biological control n.
Use of a natural enemy to
destroy a pest.
biomass n.
Dry-weight of organic tissues.
biotic a.
Those environmental
influences resulting from the
activities of living organisms.
cephalodium n.
Type of propagule in lichens
containing a blue-green alga
with a fungal mycelium.

climatic a.
Those environmental
influences resulting from
the climate.
clone n.
A group of organisms
arising by mitosis from a
single ancestor.
commensalism n.
A condition where
organisms live together
and share food.
composite a.
Type of inflorescence of
closely packed flowers found
in hawkweeds, thistles
and daisies.
crustose a.
A lichen which forms a
powdery coating on a solid.
cryptozoic a.
Fauna inhabiting dark
situations.
culm n.
The vertical stalk of a grass.

191

cymose a.
Type of inflorescence which develops from the growth of axillary buds after apical budding has ceased.

detritivore n.
An animal which feeds on waste matter (detritus).

diagnostic a.
A feature which distinguishes an organism from others similar to it.

ecological niche n.
The position in a community of a species relative to others in the community.

ecology n.
See autecology, synecology.

ecosystem n.
A situation arising from the interaction of a number of organisms with their environment and with each other.

edaphic a.
Those environmental influences resulting from soil (substrate) conditions.

elaiosome n.
An oil-body attached to a seed.

epicuticle n.
The outermost layer of the external skeleton in certain invertebrates.

epigeal a.
A plant structure which appears above the ground.

erosion n.
The wearing away of soil by physical forces.

etiolation n.
The appearance produced in plants by growth under conditions of low illumination.

exuvia n.
Structural material shed by an animal when it moults.

facultative a.
An organism which is sometimes parasitic but which can vary its mode of living.

gastropod n.
A mollusc having a single shell.

genotype n.
The genetic constitution of an animal considered in isolation from environmental influences.

geotropic a.
The type of response to gravity, by growth, of a plant organ.

glabrous a.
Descriptive of a plant structure without hairs.

gonad n.
Reproductive organ producing gametes in an animal.

guttation n.
Exudation of water and solutes from a plant.

gynomonoecious a.
A species which has female and hermaphrodite flowers on the same plant.

herbivore n.
An animal which feeds on plants.

hermaphrodite n.
An individual which produces both male and female gametes.

heterotrophic a.
Organisms taking food which is already synthesised.

hibernaculum n.
Place wherein an animal, usually poikilothermous, over-winters in a dormant condition.

histogen n.
A tissue-producing zone
in a plant.

holophytic a.
Organisms like green plants
which synthesise the whole
of their food.

holozoic a.
Organisms which ingest
solid food.

homologous a.
Structures which have a
common origin in evolution
but usually differ in
appearance.

hydathode n.
An epidermal structure,
specialized for the exudation
of water, in a plant.

hygrokinetic a.
Movements which take
place in response to
humidity changes.

hyperparasite n.
An organism parasitic on
another parasite.

inflorescence n.
A group of close-set flowers
arranged on an axis.

inquiline n.
An organism developing
within a plant gall which
another organism has
induced.

interspecific a.
Those relationships
occurring between members
of two or more species.

intraspecific a.
Those relationships
occurring between members
of a single species.

isidium n.
A relatively large propagule
fragmented from the
vegetative thallus of a lichen.

kinesis n.
A non-orientated response

by an animal to an external
stimulus.

klinokinesis n.
The response of an animal
to an external stimulus by
an increase in the number
of its random movements.

lamellibranch n.
A mollusc having a pair of
shells.

marsupium n.
A pouch in which young
animals are carried.

morphology n.
The study of the structure
of organisms.

mycorrhiza n.
A fungal mycelium
symbiotic with the roots of a
vascular plant.

neotenous a.
Condition where an animal
is capable of reproducing
while in its larval form.

node n.
The position on a stem
where leaves arise.

obligate a.
An organism which is
restricted to one mode of
life.

ocellus n.
The simple type of eye
found in certain
invertebrates.

omnivore n.
An animal which takes both
plant and animal food.

orthokinesis n.
The response by an animal
to an external stimulus
whereby there is a change
in the level of activity.

osmo-regulation n.
The control of osmotic
pressure by differential flow
of solvents through semi-
permeable membranes.

193

papilionaceous a.
The type of flower found in
members of the pea family.

parasite n.
An organism which is
nourished by the living
tissues of another.

parasitoid a.
An organism alternately
parasitic and free-living.

parthenogenesis n.
Development of ova without
fertilization.

perianth n.
The whorls of a flower
outside the male and female
parts.

pesticide n.
A chemical used to destroy
invertebrate pests.

petal n.
One of the perianth leaves
arising between the sepals
and the male and female
parts of a flower.

petaloid a.
Any plant structures having
the appearance of petals.

phenology n.
The study of an organism's
reactions to changes in
climatic conditions.

phenotype n.
The result of the reaction
of an organism's genetical
constitution with the
environment.

physiology n.
The study of the metabolic
processes in living
organisms.

phototaxis n.
The orientated response by
an animal to the stimulus
of light.

phototropic a.
Type of response to light,
by growth, of a plant organ.

phytoplankton n.
Drifting plant life present in
the upper levels of water.

phytotoxicant n.
A poisonous substance
produced by a plant.

plankton n.
See phytoplankton,
zooplankton.

poikilothermous a.
An animal whose body
temperature varies with that
of the surroundings.

pollution n.
The result of the degradation
of the environment.

polymorphic a.
A species whose members
show structural variation.

predator n.
An animal which feeds on
other living animals.

propagule n.
Any kind of body which on
separation from an
organism, can give rise to a
new individual.

pseudotrachea n.
A minute air-channel,
without a spiracular closing
mechanism, in direct contact
with an animal's tissues.

pycnidiospore n.
A type of spore in certain
fungi which arises by the
constriction of filaments
within a flask-shaped organ
(pycnidium).

radical a.
Structures which arise at the
base of a plant near ground
level.

radula n.
A horny, file-like strip in
the fore-gut of certain
molluscs, used for rasping
vegetation.

rhizome n.
A main stem which lies
horizontally near to the
soil surface.

runner n.
See stolon.

samara n.
A dry fruit having part of
the ovary wall expanded
into a wing.

saprophage n.
An organism which feeds
on carrion.

saprophyte n.
A plant which has limited
powers of synthesis and
which feeds on decaying
matter.

saprozoite n.
An animal which feeds on
decaying matter.

sepal n.
One of the outermost
perianth leaves which
encloses a flower when in bud

slurry n.
The excretion from farm
animals.

soredium n.
A propagule in lichens which
comprises both algal and
fungal partners.

stock n.
(a) A parent plant;
(b) the main stem of a plant
when very short.

stolon n.
A creeping branch (runner)

capable of giving rise to
new individuals.

successori n.
Organisms which colonize a
plant gall after the causer
has left.

symbiosis n.
A condition where two
organisms live together in a
mutually beneficial
association.

synecology n.
The study of two or more
species in relation to a
common environment.

taxis n.
The orientated response, by
movement, on the part of an
animal to an external
stimulus.

thigmotaxis n.
A locomotor reaction, on
the part of an animal, to the
stimulus of touch.

traumatic a.
A result produced by injury.

troglodytic a.
Descriptive of organisms
which dwell in caves.

trophic a.
Pertaining to nutrition.

xerophytism n.
The condition when land
plants are specialized for dry
situations.

zooplankton n.
Drifting animal life present
in the upper layers of water.

Index

Acer pseudoplatanus, 88–89, 105
Aëdes, and survival in
    freshwater, 130
Algae, association with fungi,
    67–70; colonization of
    plates, 125–129
Amblyrhincus cristatus,
    adaptations, 58–59; in food
    chain, 74
Apanteles glomeratus, as
    primary parasite, 65
Apus apus, and less restricted
    habitat, 60–61; and concept
    of community, 64
Aquarium, succession in,
    125–129, 180
Armadillium vulgare, eggs and
    young, 9; water loss, 11
Autecology, 123–124; of *Nebra
    brevicollis*, 136–143
Autotrophs, as producers, 73

Baermann funnel, 101, 142,
    146–147
Balantidium, as a saprozoite, 66
Balanus balanoides, as food of
    Nucella lapillus, 74–75
Berlese funnel, see Tullgren
    funnel
Biological control, 169, 174–176
Biotic factors, viii
Blue tit, estimating populations
    of, 156–163
Bufo bufo, estimating
    populations of, 152–156
Bufo calamita, colonization of
    pools, 130
Buteo galapagoensis, food chain
    of, 74
Buttercup-like flowers, key to,
    54–56

Callidina symbiotica, and
    restricted habitat, 57,
    111–112
Cervus elephus scoticus, and less
    restricted habitat, 62–63
Chemical control, 169
Chirocephalus diaphanus,
    interrupted succession of,
    130
Chlorohydra viridissima,
    symbiotic association, 68
Choice chambers, 16–20,
    21, 141
Clupea harengens, food chain
    of, 74
Colonization, of freshwater
    tanks, 125–129; of pools,
    130–134; by weeds, 168
Collecting, methods of, 144–148
Communities, viii, ix; concept
    of, 64; stable, 88–89;
    animal and effects of plant
    succession, 134–136
Commensalism, examples of,
    71–72

Detritivore, 82
Detritus chain, 82
Digger wasps, ecological
    niches, 114–115
Dune formation, ecology of,
    166

Earthworms, extraction of,
    147; estimation of numbers,
    150
Ecological niche, 21, 111, in
    digger wasps, 114–115; in
    gall wasps, 115–122; in a
    closed plant community,
    112–113; in pisciculture,
    113–114

Ecological efficiency, 85–86; pyramids, 78–79
Ecosystems, viii, ix; and food chains, 74–78; climax and subclimax, 94–105
Eltonian pyramid, 79
Endoparasites, 65
Energy, ix; origin of flow, 73; flow through an ecosystem, 80–82; chains, 81–82; rate of of transfer, 82–85
Erosion, 165–166; and plant cover, 169–170

Facultative parasites, fungi as, 65
Facultative symbiosis, larvae of *Unio* and *Anodonta*, 70–71
Food chains, 74–78; in pools, 130–134; and pesticides, 173
Food webs, ix, 76–78
Fungi, and association with algae, 67–70

Gall wasps, ecological niches in, 116–123
Grassland, as a subclimax ecosystem, 96–111; effect of grazing, 105–111; effect of mowing, 97–99; invertebrate population of, 100–103
Great auk, extinction of, 181

Habitat, viii, ix, concept of 57; galls as, 66
Hawaiian goose, preservation of, 180–181
Hemiteles fulvipes, as a secondary parasite, 65
Heterotrophs, as consumers, 73

Key, to British woodlice, 21–24; to buttercup-like flowers, 54–56
Kinesis, 11
Klinokinesis, 11

Leaf hoppers, removal sampling, 150–151

Lichens, as symbiotic relationships, 67–69
Ligia oceanica, 2; longevity, 10; temperature, 14
Ligidium hypnorum, 2
Lincoln index, 15, 151, 156, 163

Man, and population explosion, 183–184; and alien species, 178–179; and overcrowding, 184–185; and destruction of climaxes, 165–167
Microhabitats, examples, 57
Myxomatosis, 169

Nebria brevicollis, effect of plant succession on, 136–143
Nucella lapillus, food of, 74–75

Oniscus asellus viii; behaviour, 10–11; distribution of, 14–15; estimation of numbers, 15–16; evolution, 1, 21; excretion, 12; identification, 1–2; predation, 10; and humidity, 11, 16–20: respiration, 12, 81; temperature, 13–14; and light, 14, 17–20; water loss, 11–12; water gain, 13
Opalina, as a saprozoite, 66
Orthokinesis, 11
Osmoregulation, 1

Palaeo-ecology, 89–94
Parasites, *Apanteles glomeratus*, *Hemiteles fulvipes* and *Pteromalus* as, 65; *Cuculus canorus* as, 66–67; *Xenocoeloma brumpti* as 66
Parasitic chain, 82
Parasitism, and close relationships, 64
Pesticides, 172–173
Philoscia muscorum, embryonic period, 9
Plastics, disposal of, 185

Index

Pollution, environmental, 171–172; lead, 185; of water 176–177

Polycirrus arenivorus, as host, 66

Populations, estimation by total counts, 149–150 by removal sampling, 150–151 by marking/recapture, 151–163; bird populations by alighting, 163–164

Porcellio dilatatus, embryonic period, 9

Porcellio scaber, pseudotracheae, 4, 9; reaction to light and humidity, 17–19

Proteus anguinus, and restricted habitat, 57–58

Pteromalus, as a tertiary parasite, 65

Pyramid, of numbers, 78–79; of energy, 81

Ranunculus acris, effect of water on distribution, 48–53

Ranunculus bulbosus, distasteful sap, 96; effect, of grazing animals, 105–111, of mowing, 97–99, of water on distribution, 48–53

Ranunculus repens, asexual reproduction, 35–44,96; dispersal of seeds, 30–34; dormancy of seed, 32–33; effect, of grazing, 107–111, of light, 43–44, of shade, 28, of water, 44–53; etiolation, 28; sexual reproduction, 28–30

Recolonization, of temporary pools, 130–134

Refuse tips, and succession, 130–134

Relative humidity, 12

Rhizobium, and symbiotic relationships, 67–68

Rhodeus sericeus, symbiotic association, 70–71

Sampling, methods, 148–152

Saprozoite, *Balantidium* and *Opalina* as, 66

Saturation deficit, 12

Succession, in freshwater tanks, 125–129; interrupted, 129–134; through silting-up, 87, towards woodland climax, 103–105; of plants and effects on animal communities, 134–136; of plants on a single species, 136–143

Symbiosis, and close relationships, 67–71

Synecology, 123–124

Thigmotaxis, 20

Traps, Longwith, 147; Tring, 147; pitfall, 147

Trichoniscus, eggs and young, 9

Trophic relationships, 73–74

Tropical fish, feeding relationships, 125–129

Tullgren funnel, 101, 142, 146–147

Walton-on-the-Naze, palaeo-ecology of, 89–94

Waste, disposal of, 185

Wistman's Wood, 94

Woodlice, key to British species, 21–24

Xenocoeloma brumpti, 60; as a parasite, 66

Zoöchlorella, symbiotic association, 68